# Chemistry

## Physical and Chemical Changes in Matter

## Expanding Science Skills Series

BY

DR. BARBARA R. SANDALL

CONSULTANTS: SCHYRLET CAMERON AND CAROLYN CRAIG





ISBN 978-1-58037-522-1

Printing No. CD-404119

Visit us at www.carsondellosa.com

Mark Twain Media, Inc., Publishers
Distributed by Carson-Dellosa Publishing Company, LLC

# Table of Contents



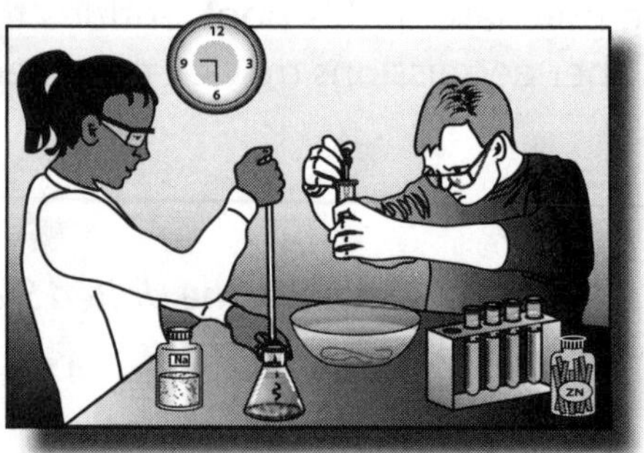

# Introduction

*Chemistry: Physical and Chemical Changes in Matter* is one of the books in Mark Twain Media's new *Expanding Science Skills Series*. The easy-to-follow format of each book facilitates planning for the diverse learning styles and skill levels of middle-school students. The teacher information page provides a quick overview of the lesson to be taught. National science and mathematics standards, concepts, and science process skills are identified and listed, simplifying lesson preparation. Materials lists for Knowledge Builder activities are included where appropriate. Strategies presented in the lesson planner section provide the teacher with alternative methods of instruction: reading exercises for concept development, hands-on activities to strengthen understanding of concepts, and investigations for inquiry learning. The challenging activities in the extended-learning section provide opportunities for students who excel to expand their learning.

*Chemistry: Physical and Chemical Changes in Matter* is written for classroom teachers, parents, and students. This book can be used as a full unit of study or as individual lessons to supplement existing textbooks or curriculum programs. This book can be used as an enhancement to what is being done in the classroom or as a tutorial at home. The procedures and content background are clearly explained in the student information pages and include activities and investigations that can be completed individually or in a group setting. Materials used in the activities are commonly found at home or in the science classroom.

The *Expanding Science Skills Series* is designed to provide students in grades 5 through 8 and beyond with many opportunities to acquire knowledge, learn skills, explore scientific phenomenon, and develop attitudes important to becoming scientifically literate. Other books in the series include *Simple Machines; Electricity and Magnetism; Geology; Meteorology; Light and Sound;* and *Astronomy.*

The books in this series support the No Child Left Behind (NCLB) Act. The series promotes student knowledge and understanding of science and mathematics concepts through the use of good scientific techniques. The content, activities, and investigations are designed to strengthen scientific literacy skills and are correlated to the National Science Education Standards (NSES) and the National Council for Teachers of Mathematics Standards (NCTM). Correlations to state, national, and Canadian provincial standards are available at www.carsondellosa.com.

# How to Use This Book

The format of *Chemistry: Physical and Chemical Changes in Matter* is specifically designed to facilitate the planning and teaching of science. Our goal is to provide teachers with strategies and suggestions on how to successfully implement each lesson in the book. Units are divided into two parts: teacher information and student information.

**Teacher Information Page**

Each unit begins with a Teacher Information page. The purpose is to provide a snapshot of the unit. It is intended to guide the teacher through the developing and implementation of the lessons in the unit of study. The Teacher Information page includes:

- National Standards: The unit is correlated with the National Science Education Standards (NSES) and the National Council of Mathematics Standards (NCTM). Pages 64–65 contain a complete list and description of the National Standards.
- Concepts/Naïve Concepts: The relevant science concepts and the commonly held student misconceptions are listed.
- Science Process Skills: The process skills associated with the unit are explained. Pages 65–68 contain a complete list and description of the Science Process Skills.
- Lesson Planner: The components of the lesson are described: directed reading, assessment, concept reinforcement through hands-on activities, materials lists of Knowledge Builder activities, and investigations.
- Extension: This activity provides opportunities for students who excel to expand their learning.
- Real World Application: The concept being taught is related to everyday life.

**Student Pages**

The Student Information pages follow the Teacher Information page. The built-in flexibility of this section accommodates a diversity of learning styles and skill levels. The format allows the teacher to begin the lesson with basic concepts and vocabulary presented in reading exercises and expand to progressively more difficult hands-on activities found on the Knowledge Builder and Inquiry Investigations pages. The Student Information pages include:

1. Student Information: introduces the concepts and essential vocabulary for the lesson in a directed reading exercise.
2. Quick Check: evaluates student comprehension of the information in the directed reading exercise.
3. Knowledge Builder: strengthens student understanding of concepts with hands-on activities.
4. Inquiry Investigation: explores concepts introduced in the directed reading exercise through labs, models, and exploration activities.

**Safety Tip:** Adult supervision is recommended for all activities, especially those where chemicals, heat sources, electricity, or sharp or breakable objects are used. Safety goggles, gloves, hot pads, and other safety equipment should be used where appropriate.

# Unit 1: Historical Perspective
## Teacher Information

**Topic:** Many individuals have contributed to the traditions of the science of chemistry.

**Standards:**
**NSES** Unifying Concepts and Processes, (F), (G)
See **National Standards** section (pages 64–65) for more information on each standard.

**Concepts:**
- Science and technology have advanced through contributions of many different people, in different cultures, at different times in history.
- Tracing the history of science can show how difficult it was for scientific innovations to break through the accepted ideas of their time to reach the conclusions we currently take for granted.

**Naïve Concepts:**
- All scientists wear lab coats.
- Scientists are totally absorbed in their research, oblivious to the world around them.
- Ideas and discoveries made by scientists from other cultures and civilizations before modern times are not relevant today.

**Science Process Skills:**
Students will be **collecting**, **recording**, and **interpreting information** while **developing the vocabulary to communicate** the results of their reading and research. Based on their findings, students will make an **inference** that many individuals have contributed to the traditions of the science of chemistry.

**Lesson Planner:**
1. Directed Reading: Introduce the concepts and essential vocabulary relating to the history of the science of chemistry using the directed reading exercise found on the Student Information pages.
2. Assessment: Evaluate student comprehension of the information in the directed reading exercise using the quiz located on the Quick Check page.
3. Concept Reinforcement: Strengthen student understanding of concepts with the activities found on the Knowledge Builder pages. **Materials Needed:** Activity #1—color pencils, construction paper, copy paper, glue, hole punch, yarn or string, scissors; Activity #2—color pencils, construction paper, glue, scissors

**Extension:** Throughout history, women have been responsible for making vast advancements in the realm of science. Students research and compile a list of 10 women scientists and their contributions to science.

**Real World Application:** Polish scientist Marie Curie's discoveries about the properties of radium paved the way for cancer therapy.

# Unit 1: Historical Perspective
## Student Information

Science as an organized body of knowledge began with the Ionian School of Greek philosophers. **Alchemy**, one of the earliest forms of chemistry, combines religion, science, philosophy, and magic. It developed in Alexandria, Egypt; China; and Greece sometime after the sixth century B.C. Archimedes (287–212 B.C.) discovered the **Law of Buoyancy** called **Archimedes' Principle**. Archimedes' Principle states that an object placed into a liquid seems to lose an amount of weight equal to the amount of fluid it displaces. Archimedes conducted an experiment to determine how much gold was in the king's crown. He did so by measuring the amount of water the crown displaced when it was submerged in water. If the crown displaced the same amount of water as an equal volume of gold, he could determine if the crown was made of pure gold.

*Archimedes*

Democritus (460–370 B.C.) developed the **Atomic Theory of Matter**, which states that all substances in the universe are made of particles that could not be broken down further. Later, these particles were called **atoms**, which is a Greek word meaning "indivisible." Democritus also explained that atoms could not be created or destroyed but could be rearranged in different combinations. This was the beginning of the development of the **Law of Conservation of Mass and Energy**.

Alchemy was the main source of chemical knowledge until 1600. Some of the discoveries made during this time included producing chemical changes in natural substances, improving methods for taking metal from ores, making and using acids, and designing balances and crucibles.

Ar-Razi (A.D. 880–909) was the first to classify chemical substances into mineral, vegetable, animal, and derivative groups. He also subdivided minerals into metals, spirits, salts, and stones.

*Ar-Razi*

In the 1500s, knowledge of chemistry was used to fight diseases. In the 1500s and 1600s, some alchemists were called **iatrochemists** because they had begun to study the chemical effects of medicines on the body. Philippus Paracelsus accepted the belief that the four basic substances were air, fire, water, and earth. He believed that the four basic substances were made of mercury, sulfur, and salt.

Libavius, who was a follower of Paracelsus, wrote the first accurate chemistry book called *Alchemia* in 1597. Jan Baptista van Helmont believed only air and water were elements, and water was the basic element of all plants. He invented the word *gas* and studied gases released by burning charcoal and fermenting wine. In 1592, Galileo developed a **thermoscope**, a precursor to the thermometer. By the 1600s, chemistry became a science. Jean Beguin wrote the first textbook of chemistry in 1611.

In the thirteenth century, Roger Bacon had begun to use the experimental method of chemical research by planning his experiments and carefully interpreting his results. Robert Boyle (1627–1691) also believed that theory must be supported by experimentation. Boyle continued Van Helmont's study of gases, and through his experiments, found that air, earth, fire, and water were not elements. The publication of his book, *The Sceptical Chymist* (1661), was the beginning of the end of alchemy. In 1662, Boyle discovered that there is an inverse relationship between the volume of gas and its pressure, now referred to as **Boyle's Law**. Boyle also rejected the current thought that matter was made of earth, air, water, and fire. He proposed that matter consisted of primary particles that could collect together to make what he called "corpuscles."

*Boyle*

*Priestly*

During the 1700s, many elements were discovered, including oxygen and its role in chemical reactions. This was one of the keys to modern chemistry. Joseph Priestly (1733–1804) conducted research on gases and discovered what would later be called **oxygen**. He found that materials burned readily in oxygen, and it had an invigorating effect if it was inhaled. He also discovered what we now know as **carbon dioxide**. While living next door to a brewery, he discovered that the fermentation of grain gave off a gas that was heavier than air and put out fire. He also discovered that when it was mixed with water, it made a refreshing drink, soda water, which was the precursor to present-day soft drinks.

Antoine-Laurent Lavoisier (1743–1794) is considered the founder of modern chemistry because of his strict approach to research. He drew up the first rational system of chemical nomenclature. He also studied combustion, and when he heard of the gas that encouraged the burning process, he called it oxygen. He defined burning as the uniting of a substance with oxygen.

*Lavoisier*

During the 1800s, fifty elements were discovered. Sir Humphrey Davy discovered sodium and potassium by running electricity through substances containing them. This process was called **electrolysis**. He also experimented with gases and discovered nitrous oxide and its properties; however, he is most well known for inventing a safety lamp for miners.

Friedrich Wöhler's (1800–1882) research developed the concepts of organic and inorganic chemistry. He and Justus von Liebig discovered that the spatial organization of atoms within a molecule was important in determining the kind of substance it made.

Chemistry was later divided into three main branches: inorganic, organic, and physical chemistry. **Inorganic chemistry** is the study of compounds without carbon. **Organic chemistry** is the study of substances containing carbon. **Physical chemistry** deals with the study of heat, electricity, and other forms of energy in chemical processes.

In 1808, John Dalton published an atomic theory suggesting that each element was made up of a certain kind of atom, and each was different from all other elements. His atomic weights were not correct; however, he did formulate the Atomic Theory of Matter. The **Atomic Theory of Matter** states that all matter is made up of atoms. His theories were based on three propositions: (1) All matter is made of extremely small indivisible particles called atoms; (2) Atoms of one element are exactly alike; and (3) When elements combine, they form compounds—their atoms combine in simple numerical proportions.

*Dalton*

In 1828, Jons Berzelius calculated more accurate atomic weights based on Dalton's atomic theory and Joseph Louis Gay-Lussac's (1778–1850) **Law of Combining Volumes**. This law states that elements combine in definite proportions by volume to form compounds. Berzelius also introduced the use of **atomic symbols**.

*Avogadro*

Amedeo Avogadro in 1811 discovered that there was a difference between atoms and molecules. Stanislao Cannizzaro demonstrated how Avogadro's theory applied to the measurement of atomic weights. This work led to the Periodic Law developed by Dmitri Mendeleev and Lothar Meyer in 1869. The **Periodic Law** states that an element's properties depend upon its atomic weight. Mendeleev developed this discovery into the periodic table of the 63 elements known during his time. He left gaps in the periodic table to show that there were still more elements to be discovered. There are currently 118 known elements.

In the 1900s, research was being done on the structure of the atom. Niels Bohr (1885–1962) proposed the first model of the atom to incorporate quantum physics. Bohr devised the concept of having the electrons in different energy levels in an atom.

*Bohr*

Name: ______________________ Date: ______________________

# Quick Check

## Matching

b 1. Archimedes
c 2. Democritus
e 3. Jean Beguin
d 4. Antoine-Laurent Lavoisier
a 5. Dmitri Mendeleev

a. developed Periodic Law
b. discovered Law of Buoyancy
c. developed Atomic Theory of Matter
d. founder of modern chemistry
e. wrote first chemistry textbook

## Fill in the Blanks

6. The Atomic Theory of Matter states that all matter is made up of atoms.
7. Organic Chemistry is the study of substances containing carbon.
8. Inorganic chemistry is the study of compounds without Carbon.
9. Joseph Priestly conducted research on gases and discovered what would later be called oxygen.
10. Niels Bohr proposed the first model of the atom to incorporate quantum physics.

### Scientists and Accomplishments

Galileo developed the thermoscope
Jean Beguin wrote first accurate textbook of chemistry
Joseph Priestly discovers oxygen and carbon dioxide
1500
1600
1700
1800
1900
Chemistry used to fight diseases
*The Sceptical Chymist* by Robert Boyle was published
50 elements discovered

## Time Line

Use the time line above to answer the following questions.

11. In the 1500s, chemistry was used to fight diseases.
12. Between 1800 and 1900, 50 elements were discovered.
13. In 1611, Jean Beguin wrote the first accurate textbook of chemistry.
14. In 1661, Robert Boyle published the book, *The* Sceptical Chymist.
15. Oxygen and carbon dioxide were discovered by Joseph Priestly.

Name: ______________________________ Date: ____________________

# Knowledge Builder

## Activity #1: Scientist Bookmark

**Directions:** Research one of the people from the list below. Using this information, fill in the blanks on the bookmark. Cut out the bookmark and glue it to a piece of construction paper the same size as the bookmark. On a blank piece of paper the same size as the bookmark, create an illustration that represents the important contribution the person made to the science of chemistry. Cut out this illustration and glue it to the back of the bokmark. Punch a hole at the top, run yarn through the hole, and tie.

Glue a picture of your scientist here.

______________________________

(name of scientist)

**Birth date:** ______________________

**Death date:** ______________________

**Nationality:** ______________________

**Important Facts**

1. ______________________________

______________________________

______________________________

______________________________

2. ______________________________

______________________________

______________________________

______________________________

**Important Scientific Contributions**

______________________________

______________________________

______________________________

______________________________

______________________________

**Name:** ______________________

**SCIENTISTS**

- Archimedes
- Democritus
- Ar-Razi
- Philippus Paracelsus
- Libavius
- Jan Baptista van Helmont
- Galileo
- Jean Beguin
- Roger Bacon
- Robert Boyle
- Joseph Priestly
- Antoine-Laurent Lavoisier
- Sir Humphrey Davy
- Friedrich Wöhler
- Justus Von Liebig
- John Dalton
- Jons Berzelius
- Joseph Louis Gay-Lussac
- Amedeo Avogadro
- Stansislas Cannizzaro
- Dmitri Mendeleev
- Lothar Meyer
- Niels Bohr

Name: ____________________ Date: ____________________

# Knowledge Builder

## Activity #2: Time Line

**Directions:** Create a historical time line. A time line is a graphic representation of a chronological sequence of events.

Step # 1: Cut out the time line and glue it on construction paper.

Step #2: Using the information from "Historical Perspective," develop a time line in sequential order (from earliest to latest) to show the history of chemistry.

Step #3: Your time line should include the title **History of Chemistry** and 14 important people and/or events with specific dates.

Step #4: Label each date with the corresponding scientist's name and the event/contribution in an organized and legible manner. Illustrations such as drawings and pictures should also be included.

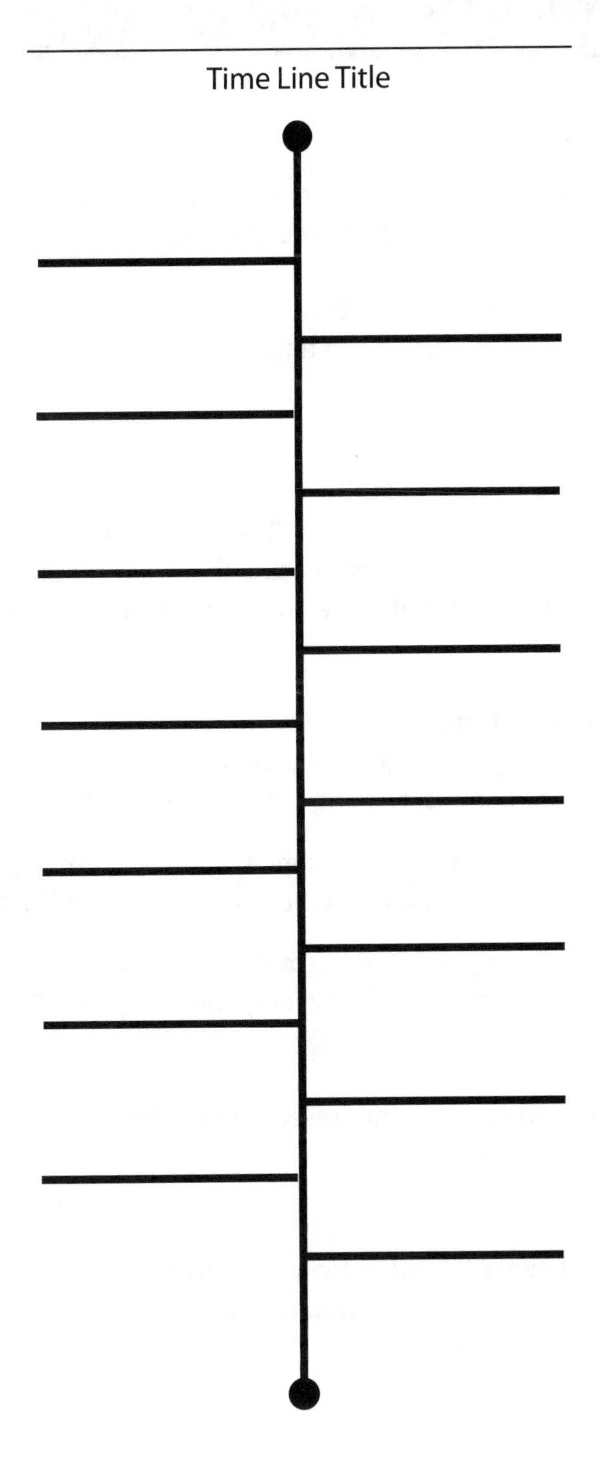

# Unit 2: Chemistry
## Teacher Information

**Topic:** Chemistry is the study of substances and how they interact with other substances.

**Standards:**
**NSES** Unifying Concepts and Processes, (B), (F), (G)
See **National Standards** section (pages 64–65) for more information on each standard.

**Concepts:**
- Chemistry is the study of substances and how they interact with other substances.
- Chemicals are not only produced by man but also in nature.

**Naïve Concepts:**
- The primary aim of chemistry is the accumulation of facts.
- Chemistry deals with artificial substances (chemicals) that are harmful and cause pollution.

**Science Process Skills:**

Students will be **collecting**, **recording**, and **interpreting information** while **developing the vocabulary to communicate** the results of their findings. Based on their findings, students will make an **inference** that there are many practical applications of chemistry in the world around them.

**Lesson Planner:**
1. Directed Reading: Introduce the concepts and essential vocabulary relating to the development of chemistry using the directed reading exercise found on the Student Information page.
2. Assessment: Evaluate student comprehension of the information in the directed reading exercise using the quiz located on the Quick Check page.
3. Concept Reinforcement: Strengthen student understanding of concepts with the activities found on the Knowledge Builder page. **Materials Needed:** Activity #1—copy paper, scissors

**Extension:** Students research and create a list of practical applications of chemistry in the world around them.

**Real World Application:** Aspirin was first developed from the bark of a willow tree. Its carbon structure was finally duplicated by scientists, and now aspirin is made synthetically.

# Unit 2: Chemistry
## Student Information

The term *chemistry* was used for the first time around A.D. 400–409, and it was used in reference to changing matter. Chemistry and the chemical industry really has its roots in the kitchen—pounding grain and other foods, boiling food in pots, straining to separate solids and liquids, fermentation, etc. Salt was probably one of the first chemicals used. Salt is found in the seas and inside the earth. Salt has been used for many things, including flavoring and preserving foods, melting snow and ice, softening water, processing fabrics and leather, mummification, making pottery, and building churches. Salt was also used as a medicine in ointments, powders, and syrups. Another commonly used early chemical was sodium. It was used as a preservative, in glasses and glazes for pottery, and in cleaning textiles. Other early chemicals were plant and animal dyes.

**Chemistry** is the study of substances and how they interact with other substances. The scientific definition of chemistry is the study of the composition of matter and the changes that the matter undergoes. Chemistry is related to many areas of science including biology, geology, physiology, physics, medicine, and so on. There are many practical applications of chemistry in the world around us. Clothes are made from synthetic fibers and natural or man-made dyes. Cooking is chemistry. For example, when baking a cake, several different substances are mixed and baked, which results in a new substance.

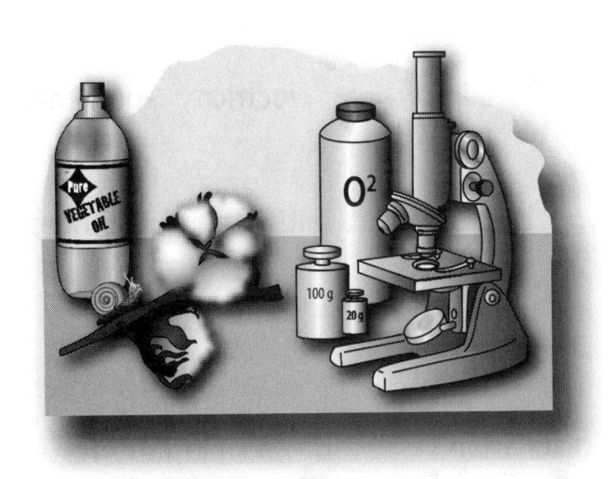


In the 1800s, chemistry was divided into three main branches: inorganic, organic, and physical chemistry. **Inorganic chemistry** is the study of compounds without carbon. **Organic chemistry** is the study of substances containing carbon. **Physical chemistry** deals with the study of heat, electricity, and other forms of energy in chemical processes. Two more branches of chemistry were added: **analytical chemistry**, which deals with the composition of substances, and **biochemistry**, which is the study of the chemistry of living organisms.

Some of the processes used by chemists are filtration, distillation, fermentation, and sublimation. **Filtration** uses porous materials to separate solids from liquids, (i.e., a coffee filter allows the coffee oils through but not the grounds). **Distillation** is a process by which a liquid is turned into a vapor and condensed back into a liquid. This process is used to separate liquids from dissolved solids or volatile liquids from less volatile ones. For example, salt can be removed from seawater by allowing the water to evaporate and re-condense in another container. **Fermentation** is the production of alcohol from sugar through the action of yeast or bacteria. **Sublimation** is when a solid turns to a gas without first changing to a liquid (i.e., mothballs).

It is important to understand the basic concepts of chemistry and its application because it is part of everyday life. One way the study of chemistry is important is in trying to change the negative effects of some by-products of industry, such as the impact that aerosol sprays have had on the ozone layer.

Name: ______________________ Date: ______________

# Quick Check

## Matching

b 1. one of the first chemicals used
a 2. study of substances with carbon
e 3. deals with the composition of substances
d 4. the study of chemistry of living organisms
c 5. study of compounds without carbon

a. organic chemistry
b. salt
c. inorganic chemistry
d. biochemistry
e. analytical chemistry

## Fill in the Blanks

6. Filtration uses porous materials to separate solids from liquids.
7. Distillation is a process by which a liquid is turned into a vapor and condensed back to a liquid.
8. Sublimation is when a solid turns to a gas without first changing to a liquid.
9. Physical chemistry deals with the study of heat, electricity, and other forms of energy in chemical processes.
10. Chemistry is the study of substances and how they interact with other substances.

## Multiple Choice

11. Making yogurt is an example of ________.
    a. filtration b. sublimation c. distillation (d.) fermentation
12. Freeze-dried substances are an example of ________.
    a. filtration (b.) sublimation c. distillation d. fermentation
13. Pure drinking water is an example of ________.
    a. filtration b. sublimation (c.) distillation d. fermentation
14. Dust collecting in a vacuum cleaner is an example of ________.
    (a.) filtration b. sublimation c. distillation d. fermentation

Name: ______________________ Date: ______________________

# Knowledge Builder

## Activity #1: Study Guide

**Directions:** Fold a sheet of white, unlined paper in half like a hotdog bun. Next, fold the paper in fourths and then in eighths. Unfold the paper. You now have a hotdog bun folded in 8 equal parts. Form 8 tabs by cutting along the folds on one side of the paper. Write the following vocabulary terms on the front tabs. Write definitions and important information under the tabs. Use the foldable to help you study chemistry words.

- Atomic Theory of Matter
- electrolysis
- Archimedes' Principle
- alchemy
- inorganic chemistry
- organic chemistry
- physical chemistry
- Periodic Law

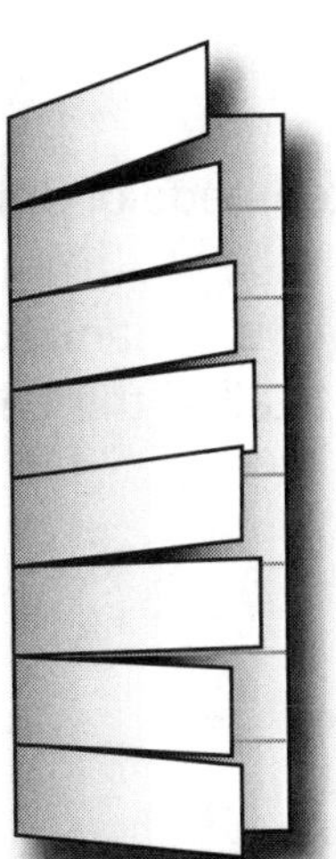

## Activity #2: Process Used in Chemistry

**Directions:** Complete the data table below. Define each process and give an example.

| Process | Definition | Example |
|---|---|---|
| Filtration | | |
| Distillation | | |
| Fermentation | | |
| Sublimation | | |

# Unit 3: Elements
## Teacher Information

**Topic:** An element is the simplest form of matter.

**Standards:**
**NSES** Unifying Concepts and Processes, (B), (G)
**NCTM** Number and Operations
See **National Standards** section (pages 64–65) for more information on each standard.

**Concepts:**
- An element is matter made of only one kind of atom.
- The list of elements is organized and displayed in a chart called the Periodic Table of Elements.

**Naïve Concepts:**
- Particles possess the same properties as the materials they compose. For example, atoms of copper are "orange and shiny," gas molecules are transparent, and solid molecules are hard.

**Science Process Skills:**

Students will be **collecting**, **recording**, and **interpreting information** while **developing the vocabulary to communicate** the results of their findings. Based on their findings, students will make an **inference** that everything on Earth is made of elements that are listed on the Periodic Table.

**Lesson Planner:**
1. Directed Reading: Introduce the concepts and essential vocabulary relating to the elements and Periodic Table using the directed reading exercise found on the Student Information pages.
2. Assessment: Evaluate student comprehension of the information in the directed reading exercise using the quiz located on the Quick Check page.
3. Concept Reinforcement: Strengthen student understanding of concepts with the activities found on the Knowledge Builder page. **Materials Needed:** Activity #1—copy paper; colored pencils; markers; a variety of small, round objects students can use for atom parts; glue; yarn or string; Activity #2—heavyweight paper, pencils, scissors

**Extension:** Students research the physical properties of two synthetic elements.

**Real World Application:** While there may be more elements in the universe to be discovered, the basic elements remain the same. Iron (Fe) atoms found on Earth are identical to iron atoms found on meteorites and on Mars.

# Unit 3: Elements
## Student Information

**Elements** are substances made up of only one kind of atom that cannot be divided by ordinary laboratory means. Ordinary laboratory means might include physical separation, filtration, or distillation. The **Periodic Law** states that an element's properties depend upon its atomic weight. Dmitri Mendeleev developed this discovery into the Periodic Table of the 63 elements known during his time. He left gaps in the Periodic Table showing that there were still more elements to be discovered. Currently there are 118 known elements. By organizing the elements by atomic number in the Periodic Table, groups of elements emerged. The horizontal rows are **periods**. The Periodic Law states that when elements are arranged by increasing atomic number, their physical and chemical properties are the same. These periods are arranged according to the atomic number, which is the number of protons in the nucleus. The columns represent groups or **families** that have similar physical and chemical properties.

Each box on the Periodic Table has the atomic number, which represents the number of protons or positively charged particles in the nucleus. The number of electrons always equals the number of protons in an electrically balanced atom. Atoms of the same element have the same number of protons but may have a different number of neutrons. Elements with different numbers of neutrons are called **isotopes. Atomic weights** are determined by comparing the element with an atom of carbon 12, which is assigned the weight of 12 units. The **atomic mass numbers** are often used in place of atomic weights. Atomic mass is the number of protons and neutrons found in the atom.

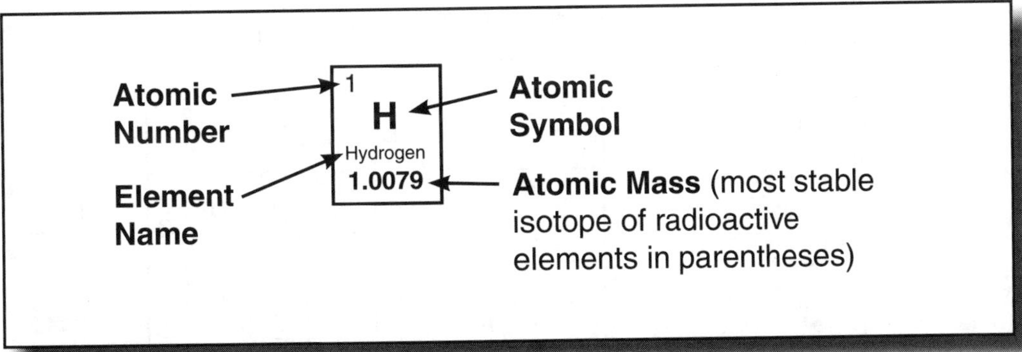


Atoms can combine to form new substances called molecules. Chemical bonds hold the atoms of the molecule together. Molecules are in constant motion. **Molecules** are the smallest part of a compound that still has the properties of the compound. A molecule is a neutral, chemically bonded group of atoms that acts as one unit. Molecules are in constant motion in all states of matter. Molecules of a solid are packed tightly together, have strong cohesive force, and move slowly. **Cohesion** is the attraction of like substances. Molecules of a liquid are spread farther apart and have a lower cohesive force that allows the molecules to slide over one another, and they move more rapidly. In a gas, molecules have very little cohesive force, are spread farther apart, and move very rapidly.

# The Periodic Table of the Elements

Metals — Nonmetals

Transition Metals

Atomic Number → 1; Atomic Symbol → H; Element Name → Hydrogen; Atomic Mass (most stable isotope of radioactive elements in parentheses) → 1.0079

| | | | | | | | | | | | | | | | | | |
|---|---|---|---|---|---|---|---|---|---|---|---|---|---|---|---|---|---|
| 1<br>**H**<br>Hydrogen<br>1.0079 | | | | | | | | | | | | | | | | | 2<br>**He**<br>Helium<br>4.003 |
| 3<br>**Li**<br>Lithium<br>6.941 | 4<br>**Be**<br>Beryllium<br>9.012 | | | | | | | | | | | 5<br>**B**<br>Boron<br>10.811 | 6<br>**C**<br>Carbon<br>12.011 | 7<br>**N**<br>Nitrogen<br>14.007 | 8<br>**O**<br>Oxygen<br>15.999 | 9<br>**F**<br>Fluorine<br>18.998 | 10<br>**Ne**<br>Neon<br>20.180 |
| 11<br>**Na**<br>Sodium<br>22.990 | 12<br>**Mg**<br>Magnesium<br>24.305 | | | | | | | | | | | 13<br>**Al**<br>Aluminum<br>26.982 | 14<br>**Si**<br>Silicon<br>28.086 | 15<br>**P**<br>Phosphorus<br>30.974 | 16<br>**S**<br>Sulfur<br>32.065 | 17<br>**Cl**<br>Chlorine<br>35.453 | 18<br>**Ar**<br>Argon<br>39.948 |
| 19<br>**K**<br>Potassium<br>39.098 | 20<br>**Ca**<br>Calcium<br>40.08 | 21<br>**Sc**<br>Scandium<br>44.956 | 22<br>**Ti**<br>Titanium<br>47.867 | 23<br>**V**<br>Vanadium<br>50.942 | 24<br>**Cr**<br>Chromium<br>51.996 | 25<br>**Mn**<br>Manganese<br>54.938 | 26<br>**Fe**<br>Iron<br>55.845 | 27<br>**Co**<br>Cobalt<br>58.933 | 28<br>**Ni**<br>Nickel<br>58.69 | 29<br>**Cu**<br>Copper<br>63.546 | 30<br>**Zn**<br>Zinc<br>65.409 | 31<br>**Ga**<br>Gallium<br>69.723 | 32<br>**Ge**<br>Germanium<br>72.64 | 33<br>**As**<br>Arsenic<br>74.922 | 34<br>**Se**<br>Selenium<br>78.96 | 35<br>**Br**<br>Bromine<br>79.904 | 36<br>**Kr**<br>Krypton<br>83.80 |
| 37<br>**Rb**<br>Rubidium<br>85.47 | 38<br>**Sr**<br>Strontium<br>87.62 | 39<br>**Y**<br>Yttrium<br>88.906 | 40<br>**Zr**<br>Zirconium<br>91.224 | 41<br>**Nb**<br>Niobium<br>92.906 | 42<br>**Mo**<br>Molybdenum<br>95.94 | 43<br>**Tc**<br>Technetium<br>(98) | 44<br>**Ru**<br>Ruthenium<br>101.07 | 45<br>**Rh**<br>Rhodium<br>102.91 | 46<br>**Pd**<br>Palladium<br>106.42 | 47<br>**Ag**<br>Silver<br>107.87 | 48<br>**Cd**<br>Cadmium<br>112.41 | 49<br>**In**<br>Indium<br>114.82 | 50<br>**Sn**<br>Tin<br>118.71 | 51<br>**Sb**<br>Antimony<br>121.76 | 52<br>**Te**<br>Tellurium<br>127.60 | 53<br>**I**<br>Iodine<br>126.90 | 54<br>**Xe**<br>Xenon<br>131.29 |
| 55<br>**Cs**<br>Cesium<br>132.90 | 56<br>**Ba**<br>Barium<br>137.33 | ♦ 57-71<br>Lanthanide series<br>(rare earth elements) | 72<br>**Hf**<br>Hafnium<br>178.49 | 73<br>**Ta**<br>Tantalum<br>180.95 | 74<br>**W**<br>Tungsten<br>183.84 | 75<br>**Re**<br>Rhenium<br>186.21 | 76<br>**Os**<br>Osmium<br>190.23 | 77<br>**Ir**<br>Iridium<br>192.22 | 78<br>**Pt**<br>Platinum<br>195.08 | 79<br>**Au**<br>Gold<br>196.97 | 80<br>**Hg**<br>Mercury<br>200.59 | 81<br>**Tl**<br>Thallium<br>204.38 | 82<br>**Pb**<br>Lead<br>207.2 | 83<br>**Bi**<br>Bismuth<br>208.98 | 84<br>**Po**<br>Polonium<br>(209) | 85<br>**At**<br>Astatine<br>(210) | 86<br>**Rn**<br>Radon<br>(222) |
| 87<br>**Fr**<br>Francium<br>(223) | 88<br>**Ra**<br>Radium<br>(226) | ◊ 89-103<br>Actinide series<br>(radioactive earth elements) | 104<br>**Rf**<br>Rutherfordium<br>(261) | 105<br>**Db**<br>Dubnium<br>(262) | 106<br>**Sg**<br>Seaborgium<br>(266) | 107<br>**Bh**<br>Bohrium<br>(264) | 108<br>**Hs**<br>Hassium<br>(277) | 109<br>**Mt**<br>Meitnerium<br>(268) | 110<br>**Ds**<br>Darmstadtium<br>(281) | 111<br>**Rg**<br>Roentgenium<br>(272) | 112<br>**Uub**<br>Ununbium<br>(285) | 113<br>**Uut**<br>Ununtrium<br>(284) | 114<br>**Uuq**<br>Ununquadium<br>(289) | 115<br>**Uup**<br>Ununpentium<br>(288) | 116<br>**Uuh**<br>Ununhexium<br>(292) | 117<br>**Uus**<br>Ununseptium | 118<br>**Uuo**<br>Ununoctium<br>(294) |

| | | | | | | | | | | | | | | | |
|---|---|---|---|---|---|---|---|---|---|---|---|---|---|---|---|
| ♦ | 57<br>**La**<br>Lanthanum<br>138.91 | 58<br>**Ce**<br>Cerium<br>140.12 | 59<br>**Pr**<br>Praseodymium<br>140.91 | 60<br>**Nd**<br>Neodymium<br>144.24 | 61<br>**Pm**<br>Promethium<br>(145) | 62<br>**Sm**<br>Samarium<br>150.36 | 63<br>**Eu**<br>Europium<br>151.96 | 64<br>**Gd**<br>Gadolinium<br>157.25 | 65<br>**Tb**<br>Terbium<br>158.92 | 66<br>**Dy**<br>Dysprosium<br>162.50 | 67<br>**Ho**<br>Holmium<br>164.93 | 68<br>**Er**<br>Erbium<br>167.26 | 69<br>**Tm**<br>Thulium<br>168.93 | 70<br>**Yb**<br>Ytterbium<br>173.04 | 71<br>**Lu**<br>Lutetium<br>174.97 |
| ◊ | 89<br>**Ac**<br>Actinium<br>(227) | 90<br>**Th**<br>Thorium<br>232.04 | 91<br>**Pa**<br>Protactinium<br>231.04 | 92<br>**U**<br>Uranium<br>238.03 | 93<br>**Np**<br>Neptunium<br>(237) | 94<br>**Pu**<br>Plutonium<br>(244) | 95<br>**Am**<br>Americium<br>(243) | 96<br>**Cm**<br>Curium<br>(247) | 97<br>**Bk**<br>Berkelium<br>(247) | 98<br>**Cf**<br>Californium<br>(251) | 99<br>**Es**<br>Einsteinium<br>(252) | 100<br>**Fm**<br>Fermium<br>(257) | 101<br>**Md**<br>Mendelevium<br>(258) | 102<br>**No**<br>Nobelium<br>(259) | 103<br>**Lr**<br>Lawrencium<br>(262) |

**Chemical properties** allow substances to chemically react to other substances to form new substances. These changes occur at the atomic or molecular level. **Atomic Theory** states that matter is made of atoms. Atoms are the smallest part of an element and are the building blocks of all matter; they combine to form elements and molecules. Atoms consist of electrons, protons, and neutrons. **Electrons** have a negative charge and circle around the nucleus of the atom. The nucleus contains protons and neutrons. **Protons** have a positive charge, and **neutrons** are neutral or have no charge. Most of the mass of an atom is from the protons and neutrons and is in the nucleus.

Models of the atom are changing as more is learned about them.

- John Dalton's (1766–1844) Atomic Theory stated that atoms were solid, indivisible mass.
- J. J. Thomson (1856–1940) discovered atoms contained electrons. He described the "plum pudding" model of an atom with charged electrons stuck into a lump of positively charged material (i.e., a ball of peanut brittle with the candy part making up the positively charged material and the peanuts as the electrons). However, this model did not describe the number of electrons and protons, their arrangement, or that electrons could be removed to form ions.
- Ernest Rutherford (1871–1937) discovered that atoms contained a nucleus. He proposed that atoms had a nucleus surrounded by electrons. He thought the rest of the atom was empty space.
- Niels Bohr (1885–1962) suggested that the electrons moved around the nucleus in concentric circular paths or orbits. He further stated that electrons in a particular path have a fixed energy. In order for them to move from one orbit to another, they must gain or lose energy. A **quantum of energy** is the amount of energy needed to move an electron from its current level to the next higher level. This concept is where the term **quantum leap**, which describes an abrupt change, originates.
- James Chadwick (1891–1974) discovered that the nuclei of atoms contained neutrons that carried no charge.
- Erwin Schrodinger (1887–1961) used quantum theory to develop the **quantum mechanical model** of the atom. In this model, electrons have a restricted value, but they do not have a specified path around the nucleus. They are in a cloud around the nucleus.

Since the current theory of atomic structure consists of electrons, protons, neutrons, and hundreds of subatomic particles, Bohr's model is the easiest level to understand.

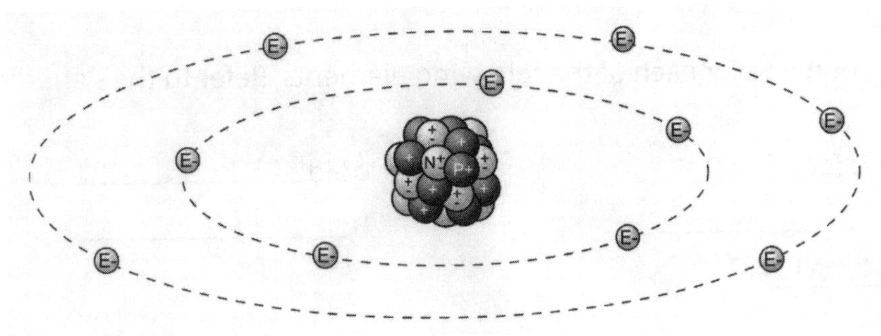

Name: ______________________ Date: ______________________

# Quick Check

### Matching

e 1. attraction of like substances

c 2. states that all matter is made of atoms

b 3. have a positive charge

d 4. have a negative charge

a 5. substances made up of only one kind of atom

a. elements
b. protons
c. Atomic Theory
d. electrons
e. cohesion

### Fill in the Blanks

6. James Chadwick discovered that the nuclei of atoms contained neutrons that carried no charge.
7. The atomic mass number are often used in place of atomic weights.
8. J. J. Thomson (1856–1940) discovered atoms contained electrons.
9. Atoms of the same element have the same number of protons but may have a different number of neutrons.
10. Currently there are 118 known elements.

### Multiple Choice

Look at the two-dimensional model of the atom to answer the questions.

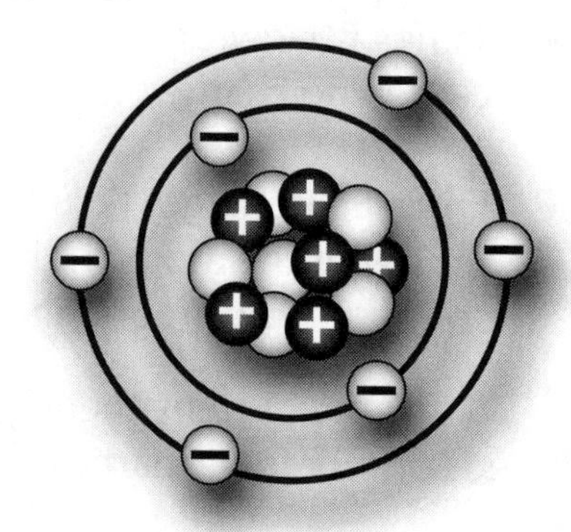

11. The atom has a protons.
    (a.) 5 b. 6 c. 7 d. 8

12. The atom has a electrons.
    (a.) 6 b. 5 c. 7 d. 4

### Reading a Table

Write the symbol or name for each of the following elements. Refer to the Periodic Table.

13. Fe Iron
14. potassium K
15. Ca Calcium
16. gold Au
17. B Boron

Name: ______________________ Date: ______________

# Knowledge Builder

## Activity #1: A 3-D Model of an Atom

**Directions:** Create a three-dimensional model of an atom that can be hung from a string. Select an element from the list below. Use the Periodic Table to determine your element's atomic number. Using the information, make a detailed sketch of the atom. Make sure to display the correct number of neutrons, electrons, and protons; these should be in their correct locations. Next, decide what to use to represent the neutrons, electrons, and protons. Anything small, round, and that can be glued to each other will work, such as ping-pong balls, small rubber balls, or Styrofoam balls. Color-code the balls so that it is easier to identify the protons, neutrons, and electrons. The electrons should be smaller than the protons and neutrons.

**Elements**

hydrogen
sodium
copper
gold
carbon
calcium

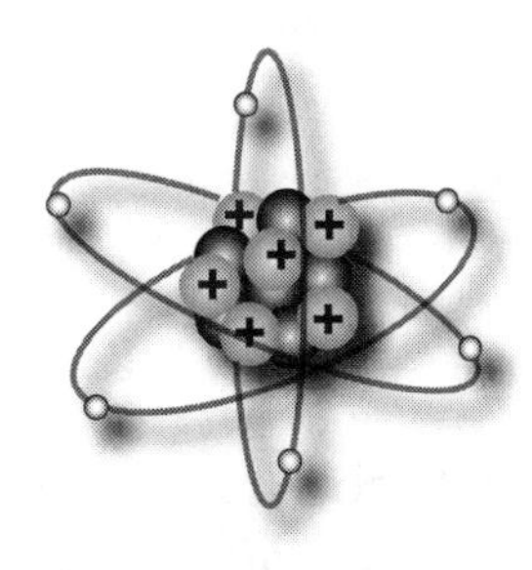

## Activity #2: Element Cube

**Directions:** The boxes on the Periodic Table provide information about the unique properties of each element. Create a cube for one of the elements on the Periodic Table. On one face of the cube, copy the information from the Periodic Table about the element you have selected. Research your element and write sentences on the other faces about the element's appearance, properties, and uses.

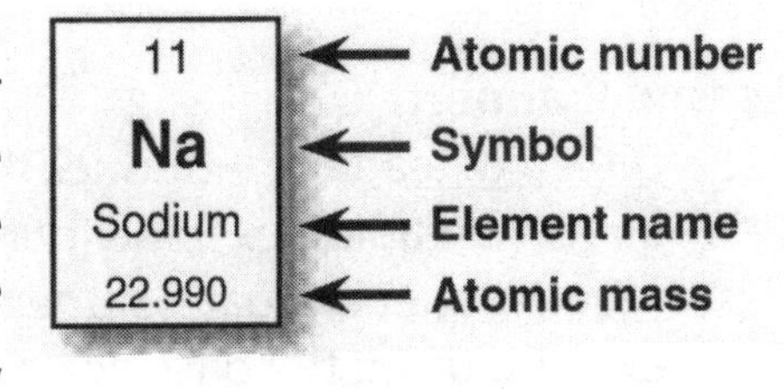

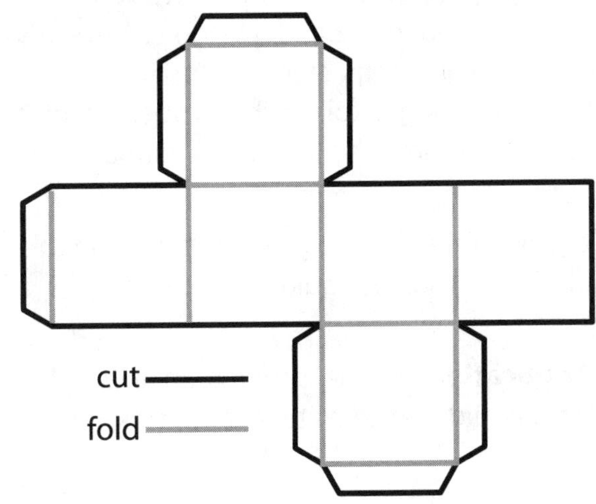

# Unit 4: The Four States of Matter
## Teacher Information

**Topic:** The four states of matter are solid, liquid, gas, and plasma.

**Standards:**

**NSES** Unifying Concepts and Processes, (A), (B), (C), (D), (E), (F), (G)

**NCTM** Data Analysis and Probability Measurement, Geometry

See **National Standards** section (pages 64–65) for more information on each standard.

**Concepts:**

- Matter is anything that has mass and takes up space.
- The four states of matter (also known as phases of matter) are solid, liquid, gas, and plasma.

**Naïve Concepts:**

- Gases are invisible and not thought of as matter.
- Gases do not have mass.
- Air and oxygen are the same gases.

**Science Process Skills:**

**Observations** will be made about materials or substances found around the classroom. The substances used will be **classified** into groups based on their common properties. The substances will be **manipulated** during the examination to determine their physical characteristics. Findings will be **communicated** through **data collections** and **writing conclusions**. Through this activity, students will develop **vocabulary** relating to the properties of matter. **Data** will be **recorded** and **interpreted** to identify possible groups for the substances.

**Lesson Planner:**

1. Directed Reading: Introduce the concepts and essential vocabulary relating to matter using the directed reading exercise found on the Student Information page.
2. Assessment: Evaluate student comprehension of the information in the directed reading exercise using the quiz located on the Quick Check page.
3. Concept Reinforcement: Strengthen student understanding of concepts with the activities found on the Knowledge Builder page. **Materials Needed:** Activity #1—heavyweight paper, old magazines or catalogs, stapler, scissors
4. Student Investigation: Explore the properties of matter. Divide the class into teams. Instruct each team to complete the Inquiry Investigation page.

**Extension:** Students take a walking field trip around the school. They identify and record substances that can be classified as solids, liquids, and gases.

**Real World Application:** Falling snow is an example of matter in the solid state. Melting snow becomes water, a liquid. As the water evaporates, it becomes water vapor, a gas.

# Unit 4: The Four States of Matter
## Student Information

**Matter** is anything that has volume (takes up space) and has mass (contains a certain amount of material). Matter doesn't have to be visible (even air is matter). All matter is made up of tiny particles that are constantly moving. The motion of the particles and the strength of attraction between the particles determine a material's state of matter. Some characteristics of matter that can be observed with your senses are color, shape, smell, taste, and texture.

There are three familiar **states of matter**: solid, liquid and gas. Plasma is a fourth state of matter that only occurs at extremely high temperatures. Plasma is like a gas, but it can conduct electricity. It is found in stars, lightning, and neon lights.

**States of Matter**

Ice Cube

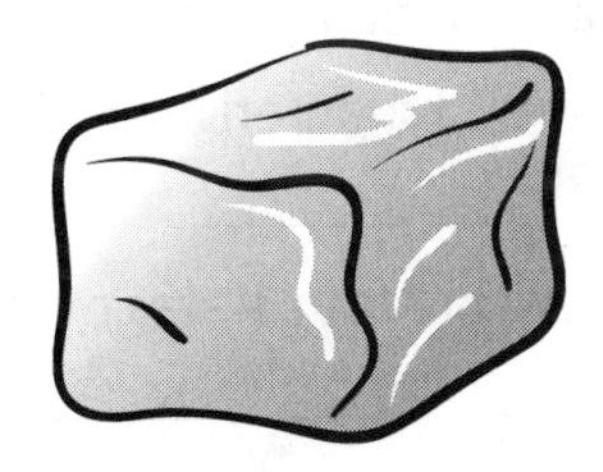

- **Solids** have a definite shape and volume. This means that solids keep their shape and take up the same amount of space.

Puddle of Water

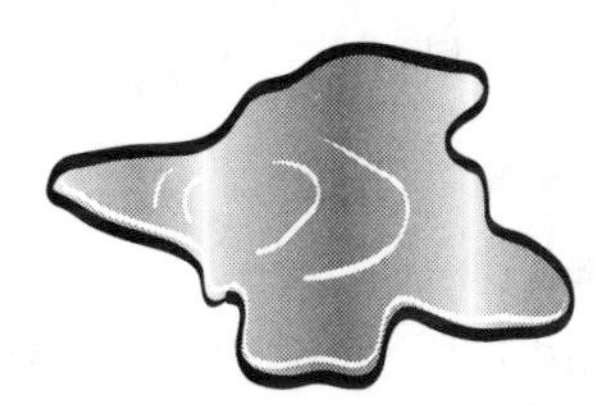

- **Liquids** have a definite volume but no definite shape. Liquids will take the shape of the container in which they are placed.

Steam

- **Gases** have no definite volume or shape. The gases will expand to fill any container and will take the shape of the container.

Solar Plasma

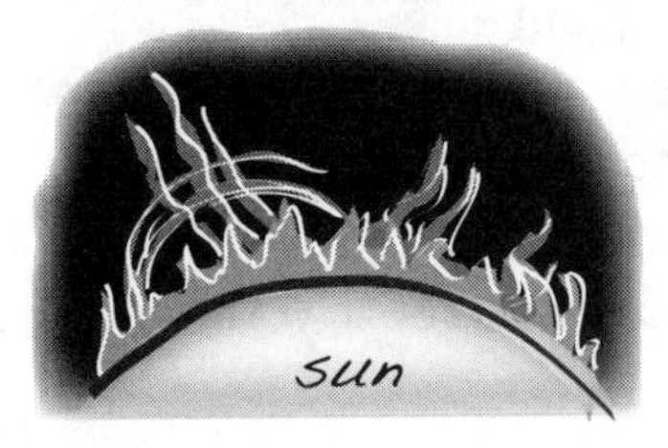


- **Plasma** has no definite shape or volume and is a highly energized gas. Plasma is common in the universe; it is not common on Earth.

Name: ______________________ Date: ______________________

# Quick Check

### Matching

__c__ 1. matter a. definite volume but no definite shape
__d__ 2. solids b. no definite volume or shape
__e__ 3. plasma c. anything that has volume and mass
__b__ 4. gases d. definite shape and volume
__a__ 5. liquids e. no definite shape or volume and is a highly energized gas

### Fill in the Blanks

6. Matter is anything that has __volume__ and __mass__.
7. Some characteristics of matter you can observe with your senses are __color__, __shape__, __smell__, __taste__, and __texture__.
8. There are four states of matter: __solid__, __liquid__, __gas__, and __plasma__.
9. __Plasma__ is a fourth state of matter that only occurs at extremely high temperatures.
10. Plasma is like a __gas__ and is found in stars, lightning, and neon lights.

### Data Table

Study the pictures and descriptions. Write the names of each picture under the correct heading.

| Solids | Liquids | Gases |
|---|---|---|
| 11. ice | 13. water | 15. air |
| 12. fish | 14. milk | 16. exhaust |

### Venn Diagram

Complete the Venn diagram by comparing matter in the solid and liquid state. You may want to copy this on your own paper if you need more room.

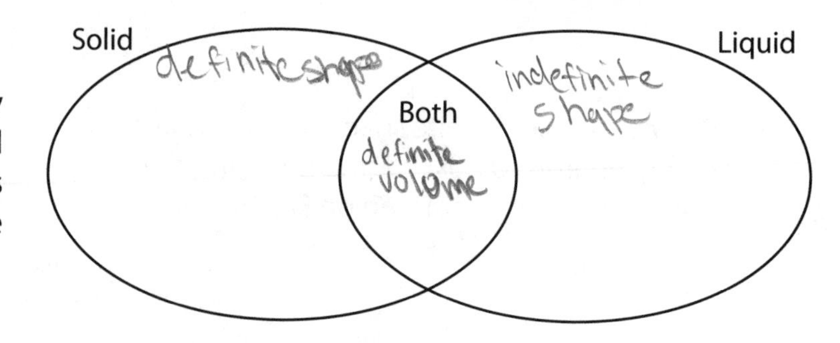

Name: ______________________ Date: ______________

# Knowledge Builder

## Activity #1: Identifying States of Matter

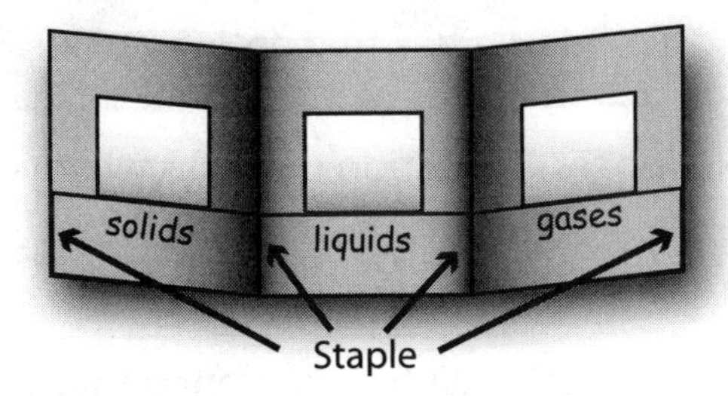


**Directions:** Using magazines, cut out examples of solids, liquids, and gases. Construct a three-pocket folder to store the pictures. Fold a horizontal sheet of paper into thirds. Fold the bottom edge up two inches and crease. Staple the two-inch tab to create three pockets. Label each pocket.

## Activity #2: Characteristics of Matter

**Directions:** Some characteristics of matter that can be observed are color, shape, smell, taste, and texture. Examine 10 objects in the classroom. Record the physical characteristics of each substance in the data table below.

| Substance | Physical Characteristics |
|---|---|
| 1. | |
| 2. | |
| 3. | |
| 4. | |
| 5. | |
| 6. | |
| 7. | |
| 8. | |
| 9. | |
| 10. | |

Group the substances listed in the table above according to characteristics they have in common.

| Characteristic | Substances |
|---|---|
| | |
| | |
| | |
| | |
| | |
| | |

Describe the characteristics you used to group the substances.

______________________________________________

______________________________________________

Name: ______________________ Date: ______________

# Inquiry Investigation: Air

**Concept:**
- Air is a **gas** that has no definite volume or shape, but it still has mass.

**Purpose:** A question that asks what you want to learn from the investigation.

*Purpose:* Does air have weight?

**Hypothesis:** Write a sentence that predicts what you think will happen in the experiment. Your hypothesis should be clearly written. It should answer the question stated in the purpose.

*Hypothesis:* ______________________________________________

______________________________________________

**Procedure:** Carry out the investigation. This includes gathering the materials, following the step-by-step directions, and recording the data.

**Materials:**
2 balloons of equal size and shape
3, 30-cm pieces of string
1 pushpin
1 meter stick
3 pieces of tape

**Experiment:**
1. Tie one of the 30-cm strings to the center of the meter stick.
2. Blow up both the balloons so they are the same size.
3. Knot the balloons. Tie a string to each knot. Tape one balloon to each end of the meter stick.
4. Suspend the meter stick from the free string. Adjust the string so the stick is evenly balanced and hangs horizontally. Pop one of the balloons with the pin.

**Results:** Write a sentence describing what you observed when you popped the balloon.

*Observation:* ______________________________________________

______________________________________________

______________________________________________

**Conclusion:** Write a brief description of what happened in the experiment and whether or not your hypothesis was correct.

______________________________________________

______________________________________________

______________________________________________

______________________________________________

# Unit 5: The Structure of Matter
## Teacher Information

**Topic:** Matter is made up of molecules.

**Standards:**
**NSES** Unifying Concepts and Processes, (A), (B)
**NCTM** Geometry
See **National Standards** section (pages 64–65) for more information on each standard.

**Concepts:**
- Matter is anything that has mass and takes up space.
- The four states of matter (also known as phases of matter) are solid, liquid, gas, and plasma.
- Matter is made up of molecules that are in constant motion.

**Naïve Concepts:**
- Expansion of matter is due to the expansion of particles rather than to increased particle spacing.
- Particles of solids have no motion.
- Relative particle spacing among solids, liquids, and gases is incorrectly perceived and not generally related to the density of the states.

**Science Process Skills:**

Students will be **inferring** the molecular structure of each of the states of matter to **create molecular models** of solids, liquids, and gases, **using the cues given** in the descriptions of each state of matter. They will be **manipulating materials** given to create the models.

**Lesson Planner:**
1. Directed Reading: Introduce the concepts and essential vocabulary relating to matter using the directed reading exercise found on the Student Information page.
2. Assessment: Evaluate student comprehension of the information in the directed reading exercise using the quiz located on the Quick Check page.
3. Concept Reinforcement: Strengthen student understanding of concepts with the activities found on the Knowledge Builder page. **Materials Needed:** Activity #1—plastic grocery sack; Activity #2—empty 2-liter plastic bottle, balloon
4. Inquiry Investigation: Explore the molecular structure of solids, liquids and gases. Divide the class into teams. Instruct each team to complete the Inquiry Investigation page.

**Extension:** Students design and construct human models of the molecular structure of a solid, liquid, and gas. For example, mark off an area. Pack in as many people as possible to represent a solid. Fewer people that move around more represent a liquid. Only three or four people moving rapidly represent a gas.

**Real World Application:** Water cycles endlessly throughout the atmosphere, oceans, land, and life of planet Earth. Surface water, such as puddles, lakes, and oceans, evaporates from the sun's heat to become water vapor (gas) in the atmosphere. The water condenses into clouds and then falls back to Earth as precipitation (rain—liquid; snow, hail, or sleet—solid).

# Unit 5: The Structure of Matter
## Student Information

This study of chemistry begins with an examination of matter. There are two theories about the makeup of matter, atomic and molecular. **Atomic Theory** states that matter is made up of atoms. **Molecular Theory** states that matter is made of molecules. **Matter** is defined as anything that has mass and takes up space or has volume. Matter has been classified into four states or phases of matter: solid, liquid, gas, and plasma.

### States of Matter

- **Solids** have a definite shape and volume. This means that solids keep their shape and take up the same amount of space. Solids have a high cohesive force so the molecules of solids are packed tightly together. The cohesive force is the attraction of like substances. Molecules are in constant motion; however, in solids, the molecules are moving so slowly you cannot see them move.

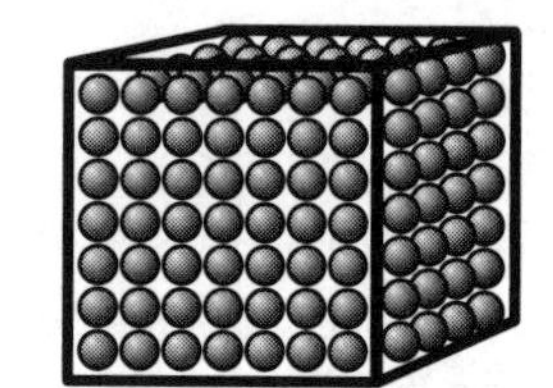

*Molecules are arranged in regular repeating patterns.*

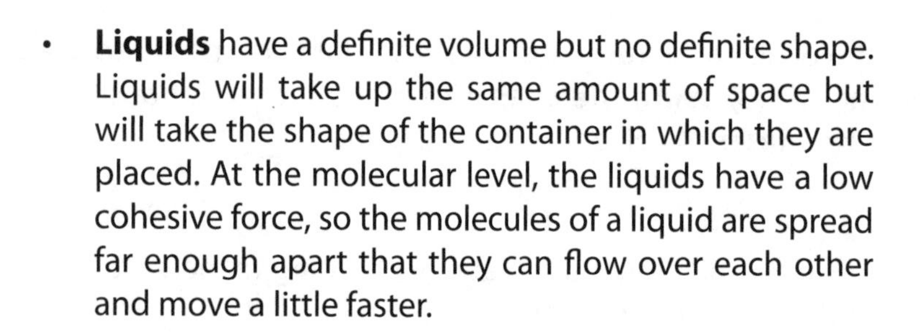

- **Liquids** have a definite volume but no definite shape. Liquids will take up the same amount of space but will take the shape of the container in which they are placed. At the molecular level, the liquids have a low cohesive force, so the molecules of a liquid are spread far enough apart that they can flow over each other and move a little faster.

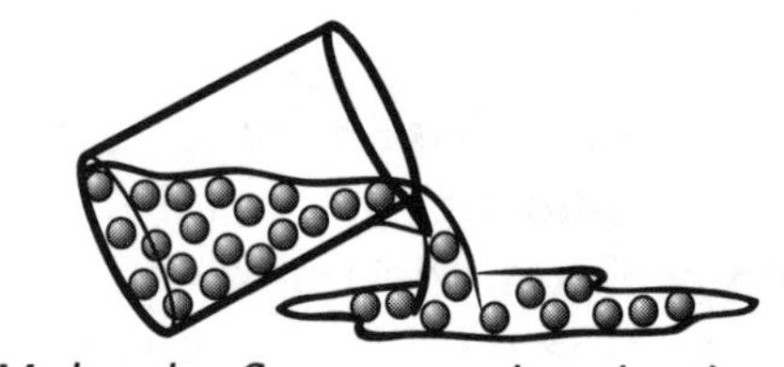

*Molecules flow around each other.*

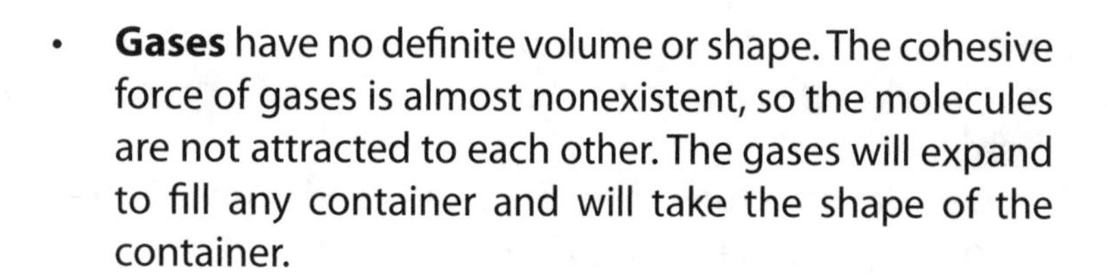

- **Gases** have no definite volume or shape. The cohesive force of gases is almost nonexistent, so the molecules are not attracted to each other. The gases will expand to fill any container and will take the shape of the container.

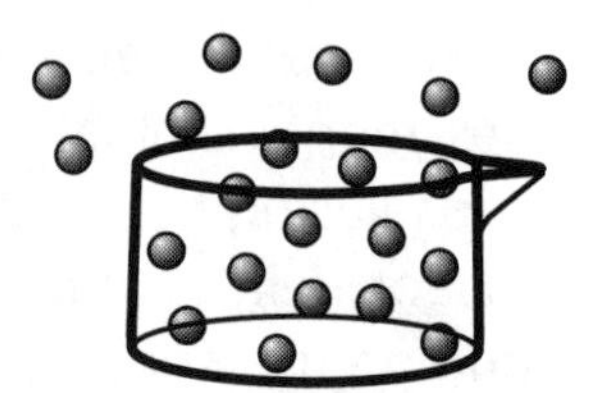

*Molecules fly around in all directions at high speeds.*

- **Plasma** has no definite shape or volume and is a highly energized gas. As with other gases, the cohesive force of plasma is almost nonexistent, so the molecules are not attracted to each other. The gases will expand to fill any container and will take the shape of the container.

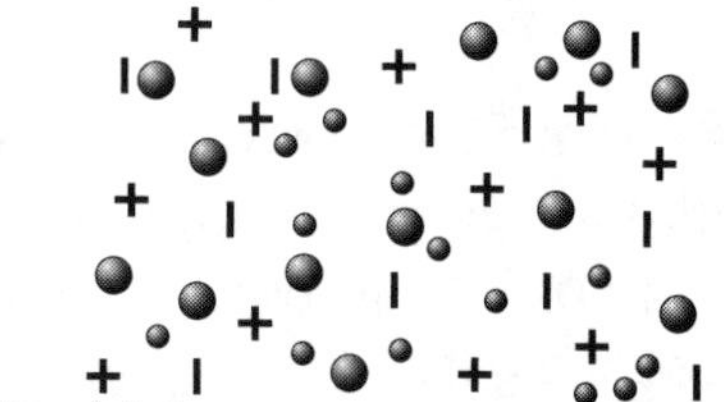

*Very high temperatures cause the atoms to lose their electrons. The result is plasma, a mixture of electrons and nuclei flying around.*

Name: ______________________ Date: ______________

# Quick Check

### Matching

| | | | |
|---|---|---|---|
| a | 1. liquids | a. | definite volume but no definite shape |
| c | 2. matter | b. | no definite volume or shape |
| b | 3. gases | c. | anything that has mass and volume |
| e | 4. plasma | d. | definite shape and volume |
| d | 5. solids | e. | no definite shape or volume and is a highly energized gas |

### Fill in the Blanks

6. Atomic Theory states that matter is made up of atoms.
7. Molecular Theory states that matter is made of molecules.
8. Matter is defined as anything that has mass and takes up space or has volume.
9. Matter has been classified into four states: liquids, Solids, gasses, and plasmas.
10. Gases will expand to fill any container and will take the shape of the container.

### Label

Each picture illustrates the movement of molecules in matter. Label the state of matter each picture represents.

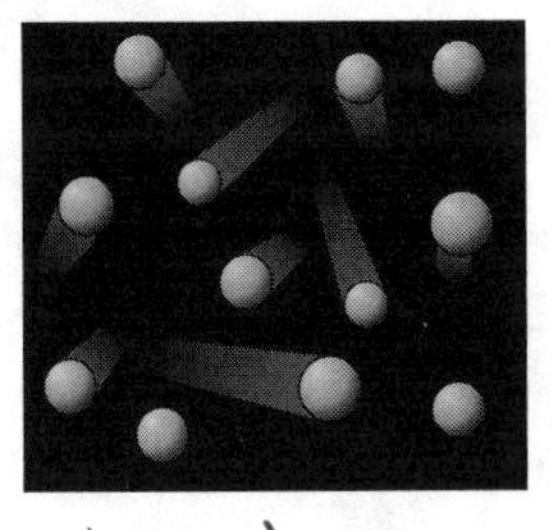

11. liquid

12. gas

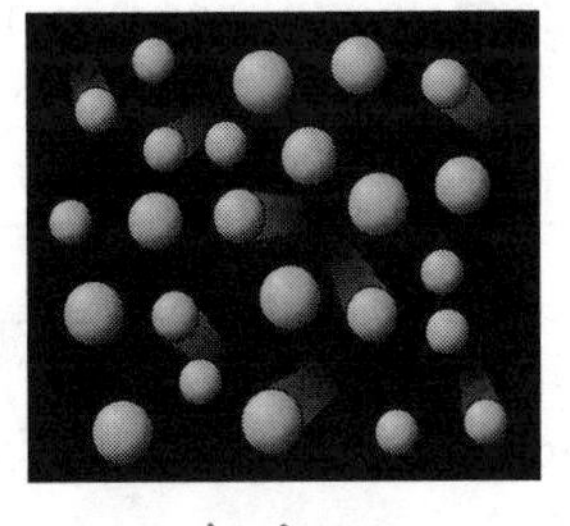

13. Solid

### Identify

The picture illustrates three states of matter. Identify the state of matter shown in the picture.

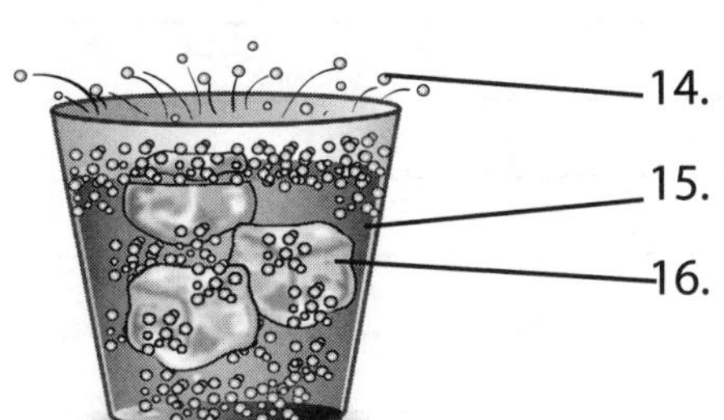

14. gas
15. liquid
16. solid

Name: ______________________ Date: ______________

# Knowledge Builder

## Activity #1: Plastic Grocery Bag

**Directions:** Take an empty plastic grocery bag. Open the top and move the bag through the air. Close the top of the bag by twisting the opening and holding it with your hand. Squeeze the bag with the other hand.

Record your observations.

______________________________

______________________________

______________________________

Conclusion: ______________________________

______________________________

______________________________

______________________________

## Activity #2: Balloon in a Bottle

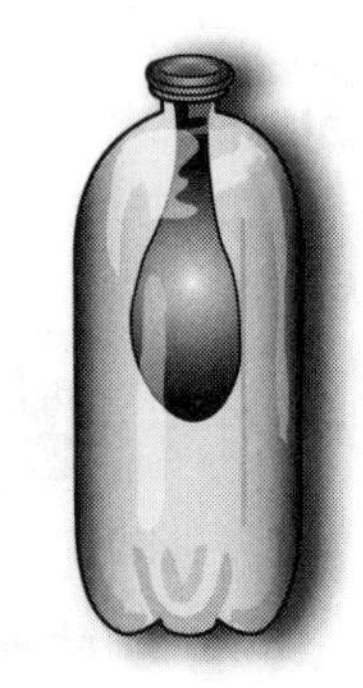

**Directions:** Hold the top of a balloon and push the bottom of the balloon inside a 2-liter bottle. Stretch the top of the balloon over the mouth of the bottle. Try to inflate the balloon by blowing into it.

Record your observations.

______________________________

______________________________

______________________________

Place a small hole in the bottom of the bottle. Try to inflate the balloon by blowing into it.

Record your observations.

______________________________

______________________________

______________________________

Conclusion: ______________________________

______________________________

______________________________

______________________________

Name: ______________________ Date: ______________

# Inquiry Investigation: Molecular Model

**Concepts:**
- **Molecular Theory** states that matter is made of particles called molecules. The state of a substance (solid, liquid, gas, or plasma) depends on how fast its particles move and how strong the attraction is between the particles.
- Water is the only known substance that can exist naturally in gaseous, liquid, and solid states of matter on Earth.

**Purpose:** Explore the molecular structure of water as a solid, liquid, and gas.

**Procedure:** Follow the directions for creating a model of the three states of water.

**Materials:**
3 plastic Petri dishes tape BBs

**Model:**
Design and construct a model of the molecular behavior of water as a solid, liquid, and gas. Use three Petri dishes, tape, and BBs to create your models.

**Results:**
Draw each of your models.

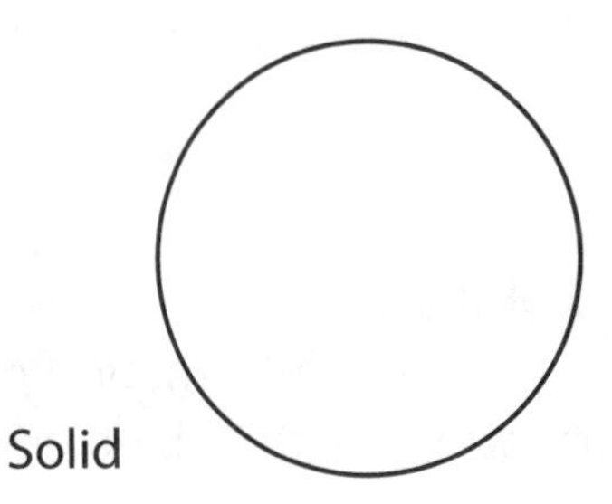


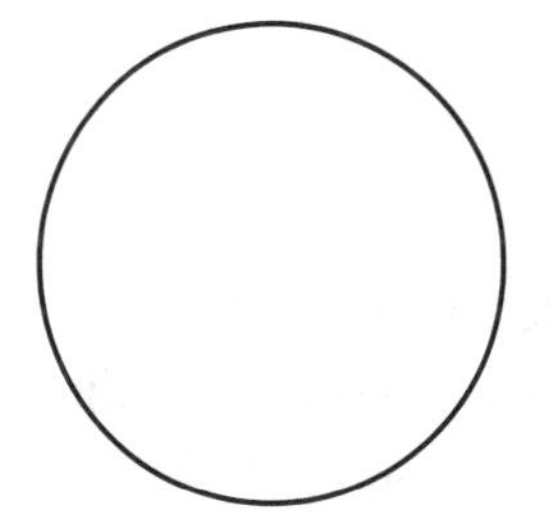


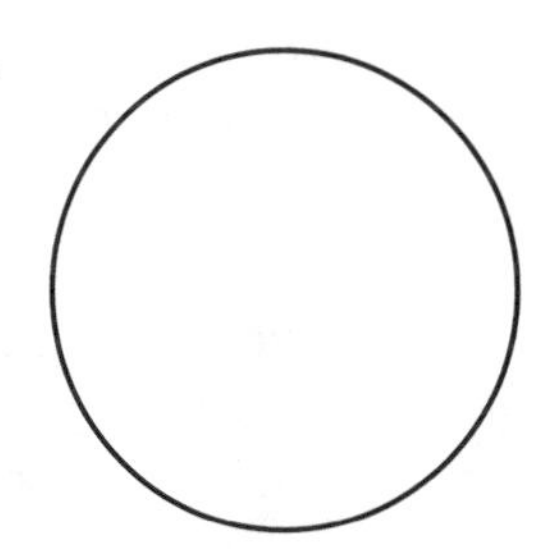

Gas

**Conclusion:**
Complete the data table below. List the characteristics of water in the solid, liquid, and gaseous states.

| State of Matter | Characteristics |
|---|---|
| Solid | |
| Liquid | |
| Gas | |

# Unit 6: Physical Properties of Matter
## Teacher Information

**Topic:** Physical properties of matter can be observed or measured without changing the chemical structure of the substance.

**Standards:**

**NSES** Unifying Concepts and Processes, (A), (B), (C)
**NCTM** Number and Operations, Measurement, Data Analysis and Probability
See **National Standards** section (pages 64–65) for more information on each standard.

**Concepts:**

- Mass is the amount of matter present. It remains constant. The metric basic unit of measure for mass is the gram.
- Volume is how much space something takes up. The metric unit of measure for volume is the liter.
- Density is the relationship between mass and volume.

**Naïve Concepts:**

- A thick liquid has a higher density than water.
- Mass and density are the same thing.
- Mass and volume are the same property.

**Science Process Skills:**

Students will be **observing**, **making predictions**, and **formulating a hypothesis and questions** about density and sinking or floating **by manipulating materials to conduct an experiment** and **using numbers to measure** volume and mass and **calculate** density. Students will be **collecting**, **recording**, and **interpreting data** while **developing the vocabulary to communicate** the results of their findings. Based on their findings, students will be able to **predict** what will sink and float in water.

**Lesson Planner:**

1. Directed Reading: Introduce the concepts and essential vocabulary related to the physical properties of matter using the directed reading exercise on the Student Information pages.
2. Assessment: Evaluate student comprehension of the information in the directed reading exercise using the quiz located on the Quick Check page.
3. Concept Reinforcement: Strengthen student understanding of concepts with the activities found on the Knowledge Builder page. **Materials Needed:** Activity #1—100 mL graduated cylinder, water, three irregularly shaped objects that will fit in the cylinder; Activity #2—three plastic cups; water; blue, red, and yellow food coloring; table salt; modeling clay; clear plastic straw, eye dropper
4. Student Investigation: Explore the physical properties of matter. Divide the class into teams. Instruct each team to complete the Inquiry Investigation pages.

**Extension:** How well a substance allows electricity to flow through it is a property of matter. Test different substances to determine if they are conductors of electricity.

**Real World Application:** Oil slicks in the ocean float on the surface because they are less dense than the salt water. Due to differences in density and cohesive forces, the oil and water do not mix.

# Unit 6: Physical Properties of Matter
## Student Information

**Physical properties** of matter can be observed or measured without changing the chemical structure of the substance. The chart below lists some common physical properties of matter.

| Physical Property | Definition | Example |
|---|---|---|
| **Density** | the amount of matter in a given volume | A rock is denser than a crumpled piece of paper of the same size. |
| **Ductility** | the ability to be stretched into a thin strand, like a wire | Metals are very ductile, and may be pulled out into wires, or hammered or rolled into thin sheets without breaking. |
| **Malleability** | the ability to be pressed or pounded into a thin sheet | Clay will bend or flatten when squeezed. |
| **Boiling Point** | the temperature at which a substance changes from a liquid to a gas | The boiling point of water is 100° C. |
| **Melting Point** | the temperature at which a substance changes from a solid to a liquid | The melting point of ice is 0° C. |
| **Electrical Conductivity** | how well a substance allows electricity to flow through it | Copper is a good conductor of electricity. |
| **Solubility** | the ability to dissolve in another substance | Sugar will dissolve in water. |

All matter can be detected and measured. Physical properties of matter that can be measured include mass, volume, and density. **Mass** is the amount of matter present, and it remains constant. The metric basic unit of measure for mass is the **gram**. Mass can be measured using a triple-beam balance. **Volume** is how much space something takes up. The volume of regular shaped objects such as a block of wood may be found mathematically. Measure the object's length, width, and height. Multiply the three measurements (L x W x H). The answer is labeled in cubic centimeters ($cm^3$).

Volume = Length x Width x Height
OR
V = L x W x H

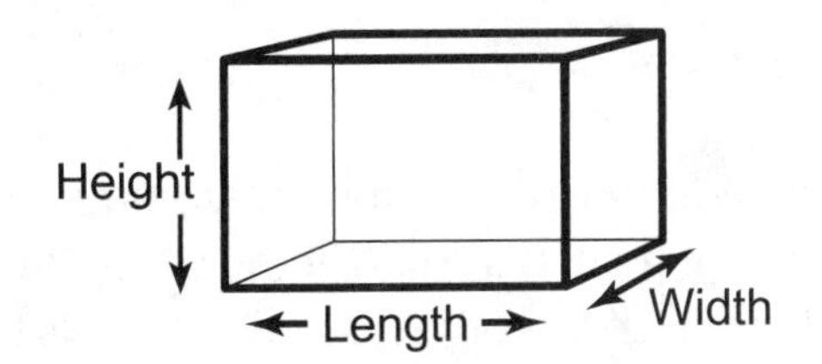

When measuring the volume of a liquid, a graduated cylinder may be used. The basic unit of measurement is the milliliter (mL) for a liquid. When reading a graduated cylinder, the liquid has a tendency to cling to the sides of the container, creating a curve called the **meniscus**. The tendency of unlike materials to be attracted to each other is called **adhesion**. When reading the volume, you must measure from the bottom of this curve.

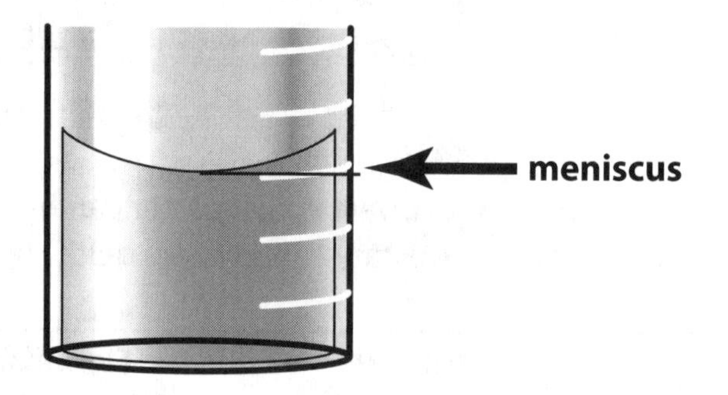


The volume of an irregular object, such as a rock, can be found by using a method called **displacement**. Water is added to a graduated cylinder and recorded. Once the volume of water has been recorded, the rock is added and the level of the water is recorded again. The difference between the first measurement and the second is the volume of the rock.

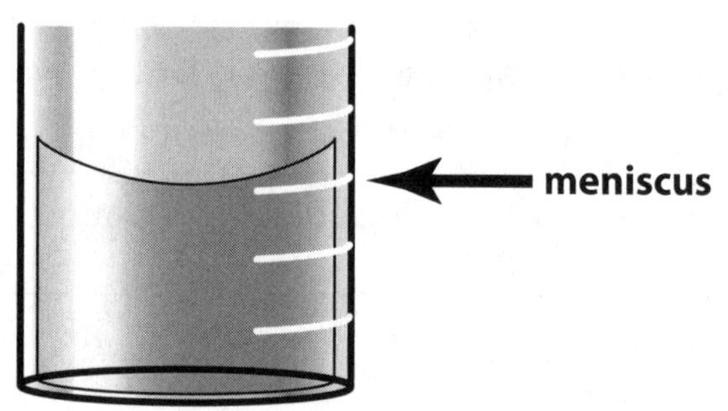


*First reading: Take reading from the bottom of the meniscus.*

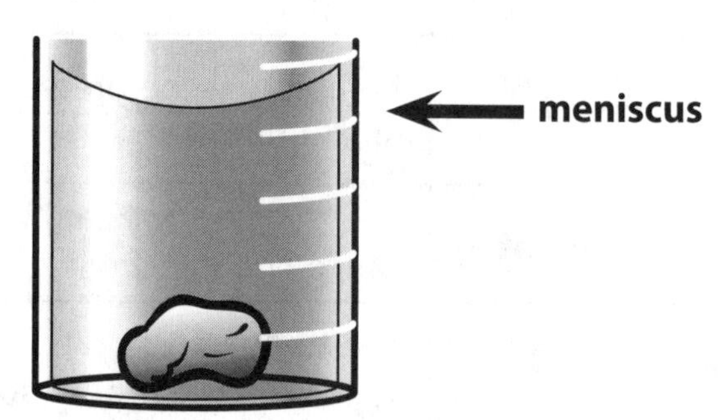


*Second reading: Take reading from the bottom of the meniscus.*

**Density** is the relationship between the mass and volume of an object. The more closely packed the molecules, the greater the density of the object. By finding the mass and volume of the object, you can find the density mathematically. It can be calculated by dividing the mass by the volume.

***Example:*** $\text{Density} = \frac{\text{Mass}}{\text{Volume}}$ or $D = \frac{M}{V}$

***Example:*** The mass of 1 milliliter of water is 1 gram.
The density of water is:

$$D = \frac{1 \text{ gram}}{1 \text{ milliliter}} \quad \text{or} \quad D = 1\text{g/mL}$$

Knowing the relationship of mass to volume or density will help you determine whether or not an object will sink or float in a liquid. If the object's density is greater than the density of the liquid it is in, the object will sink. If the object's density is less than the density of the liquid it is in, the object will float.

Name: ______________________ Date: ______________________

# Quick Check

**Matching**

| | | | |
|---|---|---|---|
| d 1. | density | a. | ability to be pounded into a thin sheet |
| e 2. | ductility | b. | allows electricity to flow through it |
| a 3. | malleability | c. | ability to dissolve in another substance |
| b 4. | electrical conductivity | d. | amount of matter in a given volume |
| c 5. | solubility | e. | ability to be stretched into a thin strand |

**Fill in the Blanks**

6. The volume of an irregular object can be found by using a method called displacement.
7. When reading a graduated cylinder, the liquid has a tendency to cling to the sides of the container, creating a curve called the meniscus.
8. Density is the relationship between mass and volume.
9. If the object's density is greater than the density of the liquid it is in, the object will sink.
10. Knowing the relationship of mass to volume or density will help you determine whether or not an object will float or sink.

**Math Calculations**

Find the volume of each three-dimensional object. Label your answers ($cm^3$).

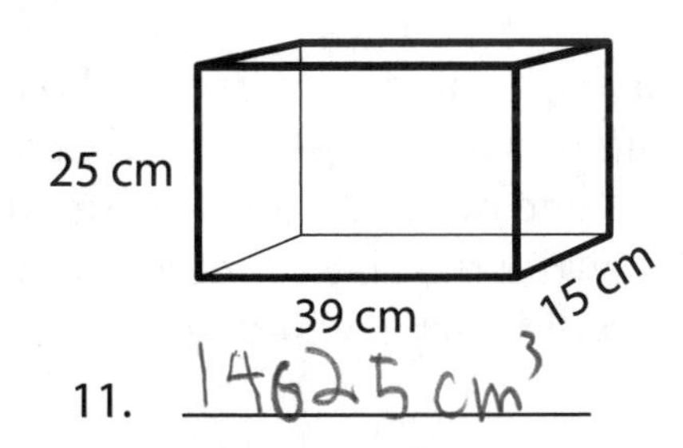


11. 14625 $cm^3$

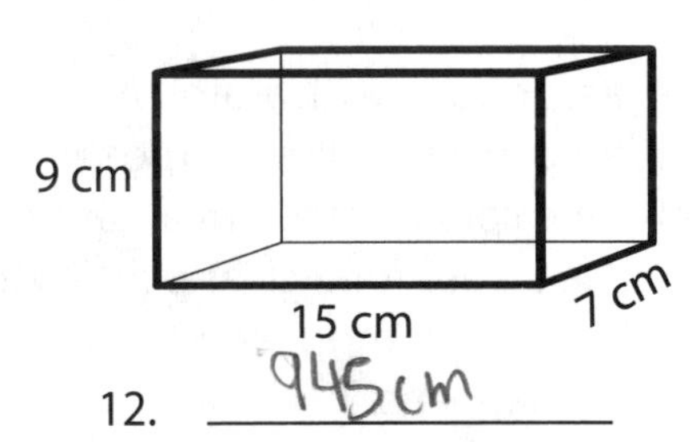


12. 945 cm

**Reading a Graduated Cylinder**

Determine the volume of the liquids in the following cylinders. Label your answers (mL).

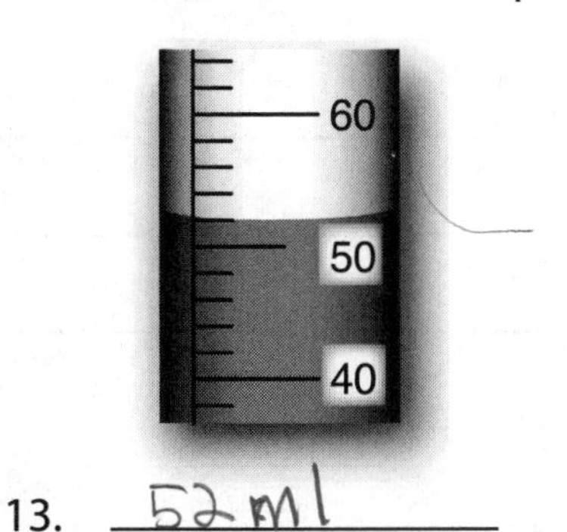


13. 52 ml

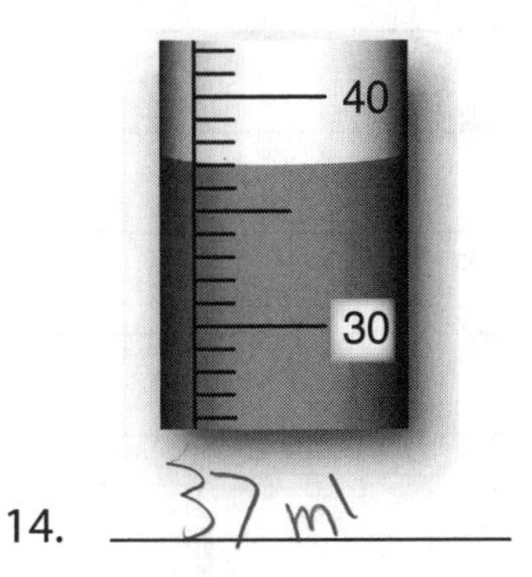


14. 37 ml

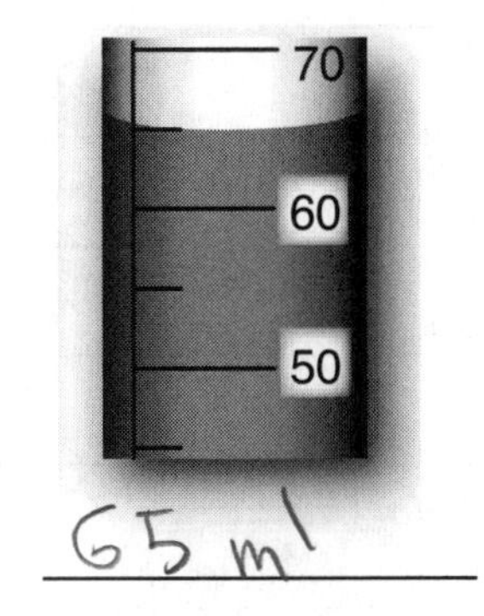


15. 65 ml

Name: ______________________________ Date: ____________________

# Knowledge Builder

## Activity #1: Finding the Volume of Irregularly Shaped Objects

**Directions:** To find the volume of irregularly shaped solid objects, use the displacement method. Collect three irregularly shaped solid objects. Fill a graduated cylinder with 50 mL of water. Record the volume in the data table. Next, lower one of the irregularly shaped objects into the water. Record the volume of the water now that the object has been added. To find the volume of the irregularly shaped object, subtract the volume of the water from the volume of the water plus the object. Record your answer in $cm^3$.

| Object | Volume of Water | Volume of Water and Object (mL) | Volume of Object ($cm^3$) |
|---|---|---|---|
| | 50 mL | | |
| | 50 mL | | |
| | 50 mL | | |

## Activity #2: Make Water Denser

**Directions:** In chemistry, the density of many substances is compared to the density of water. Does an object float on water or sink in the water? If an object such as a piece of wood floats, it is less dense than water. If an object such as a rock sinks, it is denser than water.

Fill three plastic cups with water. Add a few drops of blue food coloring to one of the cups. Add red to the other cup and yellow to the third cup. Then, add different amounts of salt to each cup. Add lots of salt to one color, a medium amount of salt to the other, and no salt to the last cup. Using modeling clay, make a base for a clear plastic straw. Stick the straw into the clay to keep it standing straight. Now, using an eye dropper, put 10 drops of each of the different colored water into the straw. Observe what happens.

Conclusion: ______________________________

______________________________

______________________________

______________________________

______________________________

______________________________

Name: ______________________ Date: ______________

# Inquiry Investigation: Density

**Concept:**
- **Density** is the relationship between the mass and volume of an object.

**Purpose:** A question that asks what you want to learn from the investigation.

*Purpose:* Does density affect an object's ability to float?

**Hypothesis:** Write a sentence that predicts what you think will happen in the experiment. Your hypothesis should be clearly written. It should answer the question stated in the purpose.

*Hypothesis:* ______________________________________________

______________________________________________

**Procedure:** Carry out the investigation. This includes gathering the materials, following the step-by-step directions, and recording the data.

**Materials:**

triple-beam balance
15 mL water
paper towels
100 mL graduated cylinder
various small objects [blocks, balls, rocks, etc. (at least one light object like a foam ball)]
masses
metric ruler

**Experiment:**

Step 1: Use the balance to find the mass of each object. Record the mass in the data table.
Step 2: Add 15 mL of water to the graduated cylinder.
Step 3: Drop an object in the graduated cylinder. Use the procedure from Knowledge Builder Activity #1 to calculate the volume of each object by subtracting the volume of just the water from the volume of the water and object. Record the volume in the data table.
Step 4: Calculate the density of each object/substance by dividing the mass by the volume.

**Results:**

Record the mass and volume for each object in the data table. Calculate and record the density of each object. The measurements for water have been recorded for you. Calculate the density and record.

| Object/Substance | Mass (g) | Volume (mL) | Density (g/mL) |
|---|---|---|---|
| Water | 15 g | 15 mL | |
| | | | |
| | | | |
| | | | |

**Analysis:** Study the results of your experiment. Decide what the data means. This information can then be used to help you draw a conclusion about what you learned in your investigation.

Create a graph. Place the mass (g) of your objects on the $y$-axis. Place the volume (mL) of your objects on the $x$-axis.

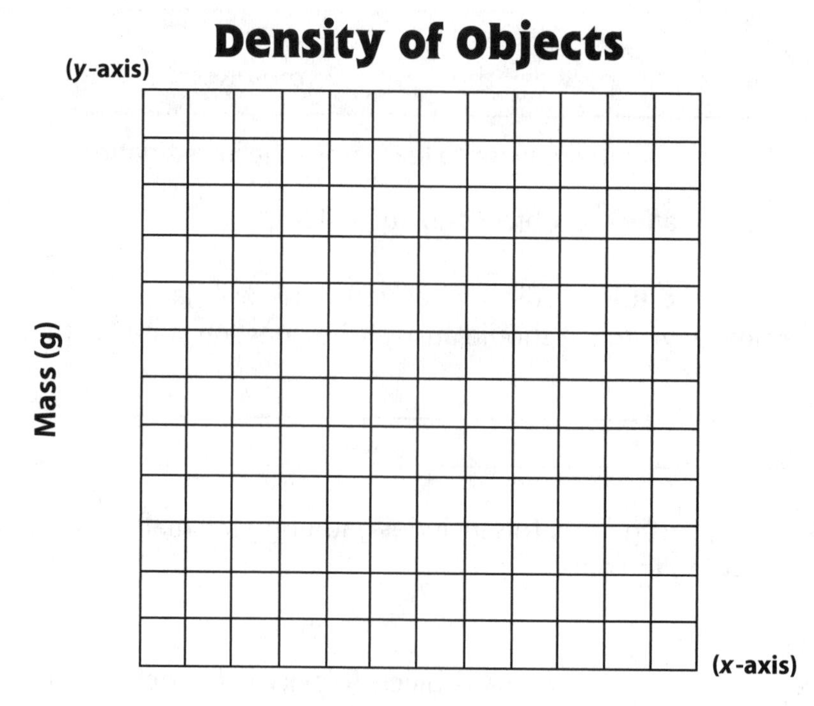


**Part II**

**Materials:**

small plastic tub filled with water

objects tested in part #1 of investigation

**Experiment:**

Step #1: Using the information illustrated by the mass and volume graph, predict and record which objects will sink and which ones will float in water.

Step #2: Record the density of each object/substance in the data table below.

Step #3: Test your predictions in the tub of water, and record the actual results.

| Object/Substance | Prediction Sink/Float | Actual Sink/Float | Why it Will Sink/Float | Density |
|---|---|---|---|---|
| | | | | |
| | | | | |
| | | | | |
| | | | | |
| | | | | |

**Conclusion:** Write a brief description of what happened in the experiment and whether or not your hypothesis was correct.

________________________________________

________________________________________

________________________________________

# Unit 7: Understanding Solids, Liquids, and Gases

## Teacher Information

**Topic:** Solids, liquids, and gases have definite characteristics.

**Standards:**

**NSES** Unifying Concepts and Processes, (A), (B), (C)
**NCTM** Number and Operations, Measurement, Data Analysis
See **National Standards** section (pages 64–65) for more information on each standard.

**Concepts:**

- Matter takes up space and has unique properties.
- The molecules of a liquid cling to each other.
- Fluid is a gas, such as air, or a liquid, such as water.

**Naïve Concepts:**

- Liquids spill and run all over.
- Solids are hard and cannot be bent or twisted.
- Liquids are the only substances classified as fluids.

**Science Process Skills:**

Students will be **observing**, **making predictions**, and **formulating a hypothesis and questions** about the characteristics of solids and liquids **by manipulating materials to conduct an experiment** and **using numbers to measure**. Students will **collect, record**, and **interpret data** while **developing the vocabulary to communicate** the results of their findings. Based on their findings, students will make an **inference** that solids, liquids, and gases have definite characteristics.

**Lesson Planner:**

1. Directed Reading: Introduce the concepts and essential vocabulary relating to the physical properties of matter using the directed reading exercise found on the Student Information pages.
2. Assessment: Evaluate student comprehension of the information in the directed reading exercise using the quiz located on the Quick Check page.
3. Concept Reinforcement: Strengthen student understanding of concepts with the activities found on the Knowledge Builder page. **Materials Needed:** Activity #1—two beakers, water, graduated cylinder, white carnation, sharp knife, blue and red food coloring; Activity #2—three large clear plastic cups, 500 mL water, 500 mL vegetable oil, 500 mL corn syrup, stop watch, three marbles of the same size
4. Student Investigation: Explore the physical properties of matter. Divide the class into teams. Instruct each team to complete the Inquiry Investigation pages.

**Extension:** Place a penny facing heads up on a flat surface. Carefully add one drop of water at a time onto the surface of the penny. Count the number of drops added to the surface before the water spills over the edge of the penny.

**Real World Application:** Surface tension allows a water strider bug to stand on the surface of water.

# Unit 7: Understanding Solids, Liquids, and Gases
## Student Information

Solids, liquids, and gases have definite characteristics. A solid has a definite shape and volume. Solids resist being pulled apart. Another characteristic is elasticity. **Elasticity** means that a solid can be bent or twisted a limited amount, and it will return to its former shape. A rubber band is a solid that can easily be stretched, and then it will return to its original shape and size. Some solids are also **malleable**, which means the solid can be formed into a new shape. Gold is a solid that can easily be hammered into a thin sheet.

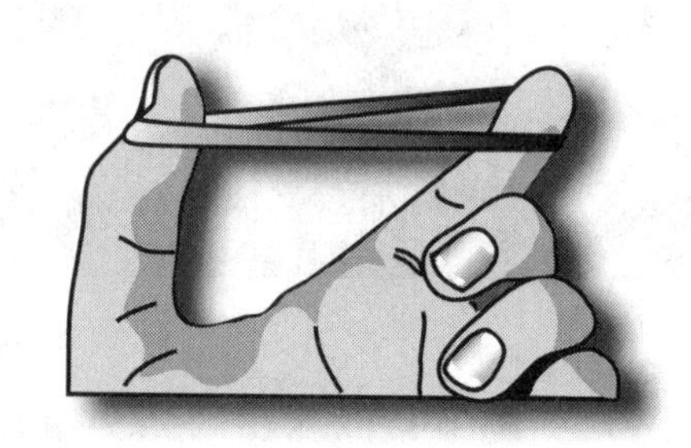

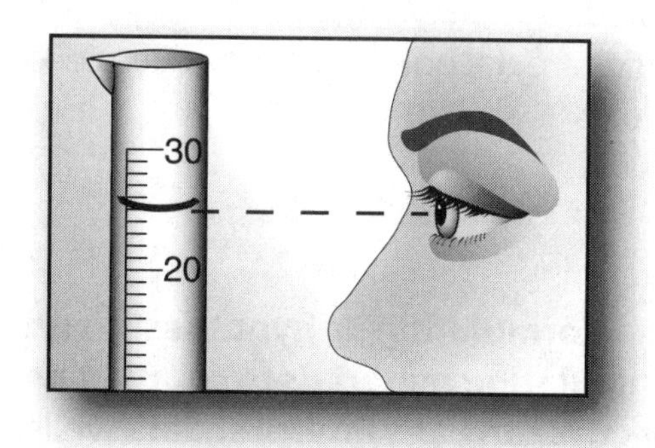


A liquid has a definite volume but takes the shape of the container holding it. When water is poured from one container into a container shaped differently, the water takes the shape of the new container. Other physical characteristics of liquids include adhesion and cohesion. **Adhesion** is the attraction between unlike substances. When measuring liquid volume, you read the measurement from the bottom of the meniscus. The meniscus is the curve formed because of the adhesion of liquid to the container. Liquids also have **capillarity**. When a thin straw or glass tube is dipped in a liquid, the liquid will rise slightly in the straw or glass tube. This concept explains why colored water seems to climb up a paper towel.

Water molecules are attracted to each other because of a force called cohesion. **Cohesion** is the attraction of like substances. As the molecules cling to each other, they form an invisible "skin" on the surface of the water or what scientists call **surface tension**. Surface tension helps a drop of water hold its shape, to hang on to itself, and stack up on a surface. It also helps some insects walk on the surface of the water.

When you pour a liquid from one container to another, the liquid takes the shape of the container. The molecules of liquids are spread far enough apart they can flow over each other and move more freely than in a solid. Some liquids flow more easily than others. A liquid's resistance to flow is known as the liquid's **viscosity**. The slower a liquid flows, the higher its viscosity. Honey has a high viscosity, while water has a lower viscosity.

Scientists describe **fluids** as any material that doesn't maintain a definite shape. Both liquids and gases are fluids. The molecules that make up these fluids do exert a pressure on surfaces with which they come in contact.

**Characteristics of Fluids**

- Fluids move from an area of high pressure to one of low pressure. ***Example:*** When you suck air out of the top of a straw, you reduce the air pressure at the top of the straw. Normal atmospheric pressure then forces liquid from your drink up the straw to your lips.

- Gravity pulls on fluids. Pressure in any contained quantity of fluid will increase with depth. ***Example:*** When holes are placed in a container, water will squirt out farthest from the hole nearest the bottom of the container.

- Water exerts an upward force on objects. The pressure pushing up on an immersed object is greater than the pressure pushing down on it. The force is known as **buoyant force**. **Archimedes' Principle** explains why objects sink or float. If the buoyant force (upward force) is equal to the weight of an object, the object will float. If the buoyant force (upward force) is less than the weight of an object, the object will sink. ***Example:*** If you jump in a tub of water, you will notice that the water exerts an upward force that keeps you from hitting bottom right away.

- **Bernoulli's Principle** reveals that the pressure in a moving stream of fluid is less than that in the surrounding fluid. ***Example:*** A plane's wings are curved on top and flat on the bottom. Air must travel farther (and faster) over the top surface than the bottom, providing a net upward force beneath the wings.

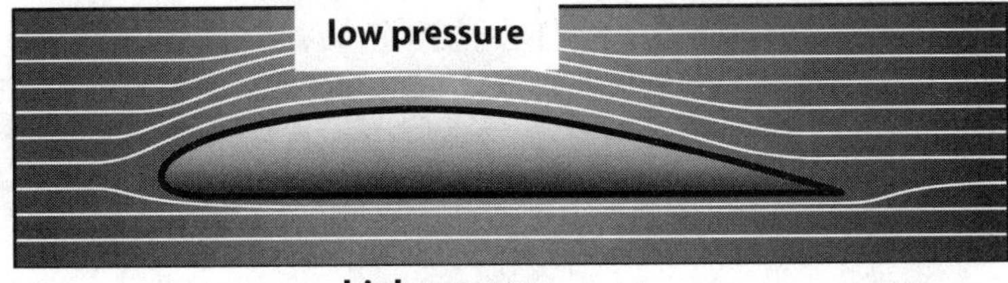

Name: ______________________________ Date: ______________

# Quick Check

## Matching

| | | | |
|---|---|---|---|
| ______ 1. | elasticity | a. | water can climb up a paper towel |
| ______ 2. | malleable | b. | a solid can be formed into a new shape |
| ______ 3. | viscosity | c. | a solid can be bent or twisted and it will return to its original shape |
| ______ 4. | capillarity | d. | a liquid's resistance to flow |
| ______ 5. | fluids | e. | liquids and gases |

## Fill in the Blanks

6. Solids, liquids, and gases have definite ______________.
7. Gold is a ______________ that can easily be hammered into a thin sheet.
8. A rubber band has ______________ and can easily be stretched.
9. The meniscus is the curve formed because of the ______________ of liquid to the container.
10. ______________ ______________ helps some insects walk on the surface of the water.
11. Water molecules are attracted to each other because of a force called ______________.
12. ______________ ______________ explains why objects sink or float.
13. ______________ ______________ reveals that the pressure in a moving stream of fluid is less than that in the surrounding fluid.

## Label

Identify the physical property illustrated in each picture below. Use the words **elasticity**, **adhesion**, **capillarity**, or **surface tension** to label the pictures.

   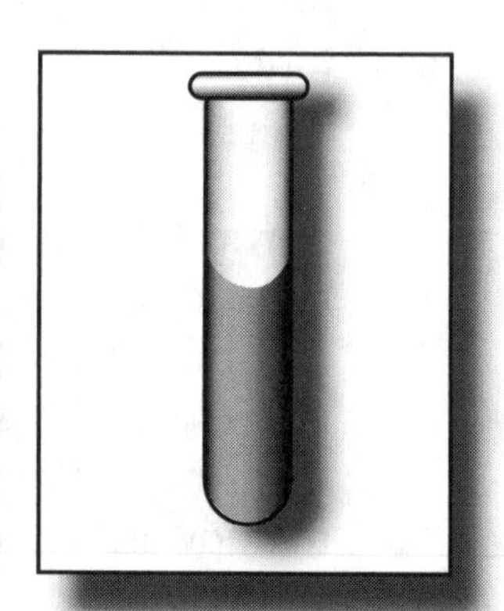

14. ______________ 15. ______________ 16. ______________ 17. ______________

Name: ______________________ Date: ______________

# Knowledge Builder

## Activity #1: Capillary Action in Plants

**Directions:** Fill 2 beakers with 250 mL of water. Add 10 drops of red food coloring to one beaker. Add 10 drops of blue food coloring to the second beaker. Slice the stem of a white carnation lengthwise, stopping 5 centimeters from the flower. Place one end of the stem in the beaker of red water and the other in the beaker of blue water. Place the flower and beakers in a sunny place. After 24 hours, observe the flower. Sketch your observations below.

Conclusion: ______________________________________________

## Activity #2: Viscosity

**Directions:** Number three large clear plastic cups. Fill cup #1 with 500 mL of water, cup #2 with 500 mL of vegetable oil, and cup #3 with 500 mL of corn syrup. Using a stop watch, time in seconds how long it takes for a marble to fall through each liquid. Record the time in the data table below.

| Liquid | Time |
|---|---|
| Water | |
| Vegetable oil | |
| Corn syrup | |

Conclusion: ______________________________________________

Name: ______________________ Date: ______________

# Inquiry Investigation: Elasticity

**Concept:**
- **Elasticity** is a characteristic of solids.

**Purpose:** A question that asks what you want to learn from the investigation.

*Purpose:* Does the temperature of a rubber band affect the distance it will stretch?

**Hypothesis:** Write a sentence that predicts what you think will happen in the experiment. Your hypothesis should be clearly written. It should answer the question stated in the purpose.

*Hypothesis:* ______________________________________________

______________________________________________

**Procedure:** Carry out the investigation. This includes gathering the materials, following the step-by-step directions, and recording the data.

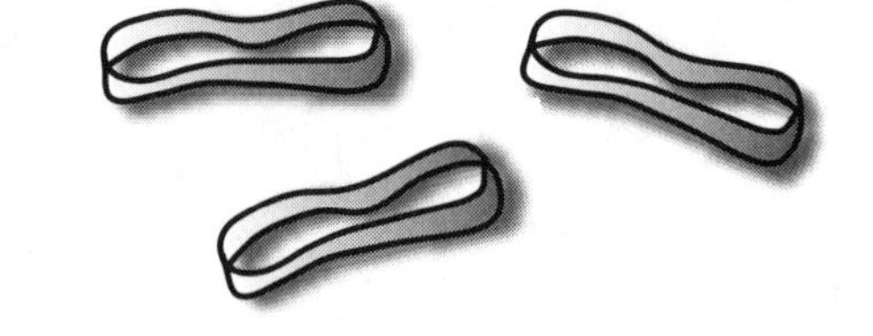

**Materials:**
10 rubber bands, the same size (5 room temperature and 5 placed in the freezer for 15 minutes)
masking tape
2 wooden clothespins
meter stick

**Experiment:**
Step 1: Secure a meter stick to the floor with tape.
Step 2: Starting at "0," lay one room-temperature rubber band alongside the meter stick.
Step 3: Have your partner grasp one end of the rubber band with a wooden clothespin.
Step 4: Grasp the other end of the rubber band with a wooden clothespin. Stretch the rubber band along the meter stick until it breaks.
Step 5: Record the distance stretched in the data table.
Step 6: Repeat the procedure with the other four room-temperature rubber bands.
Step 7: Repeat steps 2–6 for the frozen rubber bands.

**Results**
Record the distance each rubber band was stretched until broken in the data table below. Calculate and record the average for each group.

| | Elasticity | | | | | |
|---|---|---|---|---|---|---|
| **Temperature** | Trial #1 | Trial #2 | Trial #3 | Trial #4 | Trial #5 | Average |
| Room | | | | | | |
| Frozen | | | | | | |

Name: ______________________________ Date: ____________________

**Analysis:** Study the results of your experiment. Decide what the data means. This information can then be used to help you draw a conclusion about what you learned in your investigation.

Create a graph that will compare the average distance the rubber band was stretched. Place the distance (cm) on the *y*-axis. Place the temperature on the *x*-axis.

## Elasticity

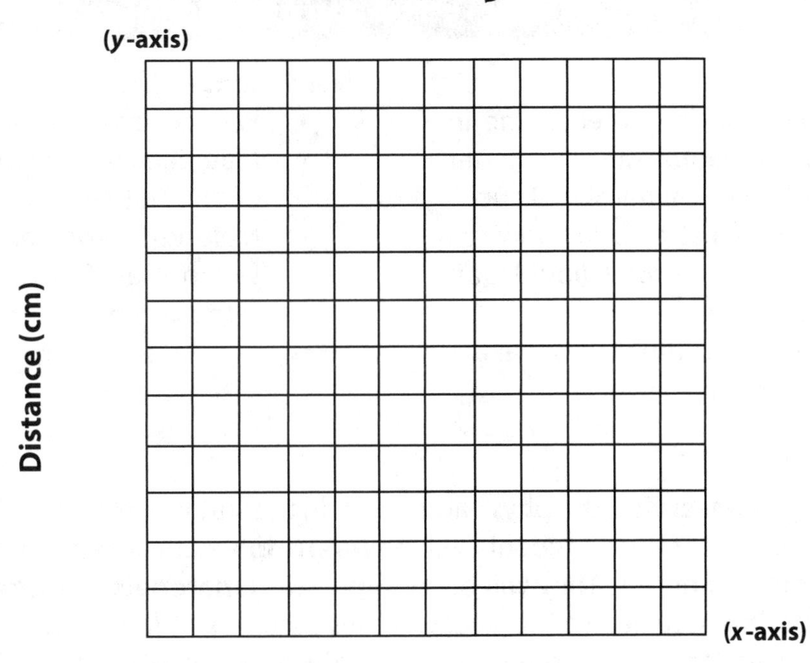


**Temperature**

**Conclusion:** Write a brief description of what happened in the experiment and whether or not your hypothesis was correct.

______________________________________________________________________

______________________________________________________________________

______________________________________________________________________

______________________________________________________________________

______________________________________________________________________

______________________________________________________________________

______________________________________________________________________

______________________________________________________________________

# Unit 8: Physical Changes of Matter
## Teacher Information

**Topic:** Matter can be changed without changing composition.

**Standards:**
**NSES** Unifying Concepts and Processes, (A), (B), (F)
**NCTM** Geometry, Number and Operations, Measurement, Data Analysis and Probability
See **National Standards** section (pages 64–65) for more information on each standard.

**Concepts:**
- A physical change in matter is one in which the form or appearance of matter changes but not its chemical makeup.
- The four states of matter (also known as phases of matter) are solid, liquid, gas, and plasma.
- Matter is made up of molecules that are in constant motion.

**Naïve Concepts:**
- Particles of solids have no motion.
- Materials can only exhibit properties of one state of matter.
- Melting/freezing and boiling/condensation are often understood only in terms of water.

**Science Process Skills:**
Students will be **observing** the physical changes in the states of matter, **using numbers**, and **measuring** time and temperature. Students will be **inferring** what caused the states of matter to change from one form to another. They will be **manipulating materials** to **make predictions** and **conducting an experiment** to determine what changes occurred. Students will be **communicating** and **developing vocabulary** during the process of **collecting, recording, analyzing,** and **interpreting data**.

**Lesson Planner:**
1. Directed Reading: Introduce the concepts and essential vocabulary related to the physical changes of matter using the directed reading exercise on the Student Information pages.
2. Assessment: Evaluate student comprehension of the information in the directed reading exercise using the quiz located on the Quick Check page.
3. Concept Reinforcement: Strengthen student understanding of concepts with the activities found on the Knowledge Builder page. **Materials Needed:** Activity #1—large balloon, glass flask, ring stand, Bunsen burner, tongs, hot pads, colored pencils; Activity #2—two cups, hot water, cold water, graduated cylinder, two tea bags, two thermometers
4. Student Investigation: Explore the effects of temperature on changes of state. Divide the class into teams. Instruct each team to complete the Inquiry Investigation pages.

**Extension:** Students explore the transfer of heat through convection. Draw a spiral on a sheet of paper and add a snake's head at the end. Cut out the spiral and suspend it by a piece of string tied to the center of the spiral over a source of heat.

**Real World Application:** Each day, about four trillion gallons of water fall to Earth as precipitation, such as rain, snow, or hail. Accurate weather predictions are important for planning our day-to-day activities.

# Unit 8: Physical Changes of Matter
## Student Information

The **states of matter** are solid, liquid, gas, and plasma. Matter can undergo different kinds of changes—physical, chemical, and nuclear. A **physical change** in matter is one in which the form or appearance of matter changes but not its chemical makeup. Whenever you cut, tear, grind, or bend matter, you are causing a physical change. The size and shape of matter has been changed without changing the chemical makeup. Tearing paper changes its physical appearance (shape and size), but it does not change its chemical makeup. It is still paper.

Physical changes include changes in the states or forms of matter. Heating and cooling can change matter from one state to another. Ice changing into water is a physical change where ice (a solid) is changed to water (a liquid). During a physical change in matter, the mass stays the same because there is not a gain or loss of matter. This is referred to as the **Law of Conservation of Mass**.

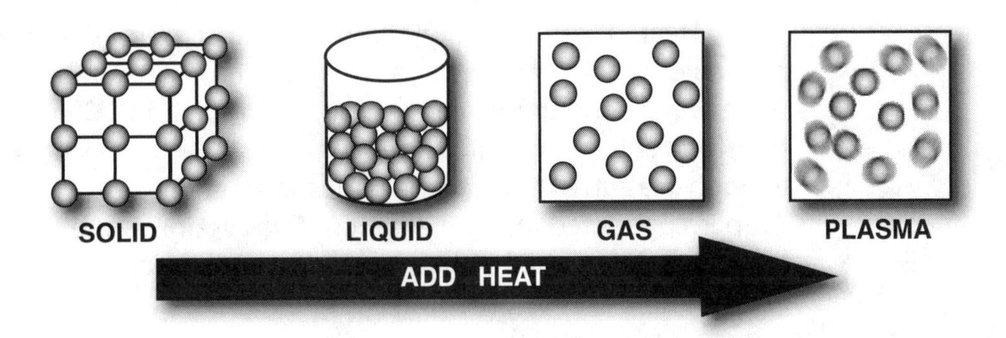


The molecules in matter are always moving. Their speed changes when they are heated or cooled. If a substance is heated or cooled enough, it changes state, or form. The molecules in a solid move back and forth, but they do not move away from each other. As a solid is heated, the molecules move faster and faster until the bonds weaken and the solid melts (changes to a liquid). The change from a solid state to a liquid state is called **melting**. The temperature at which a substance changes from a solid to a liquid is called the **melting point**. When a liquid is cooled, its molecules slow down and it may even freeze (change to a solid). The change from the liquid state to the solid state is called **freezing**. The temperature at which a substance changes from a liquid to a solid is called the **freezing point**.

Evaporation, condensation, and sublimation are processes of changing the state or form of matter by heating and cooling. **Evaporation** is the process of a liquid changing to a gas when it is heated. When a liquid is heated, the molecules move even faster. If a liquid gets hot enough, it boils and evaporates (changes to a gas). The molecules break the weak bonds holding them away from each other. When alcohol is placed on your body, it has a cooling sensation because the heat in your body causes the alcohol to quickly change into a gas. **Condensation** is the process of a gas changing to a liquid when it is cooled. When a gas is cooled enough, its molecules slow down until the gas condenses (changes to a liquid). Water vapor in the air forms clouds during condensation. When warm breath comes in contact with a cold window, the water vapor in the breath condenses on the window. **Sublimation** occurs when a solid changes directly to a gas without changing to a liquid first. Mothballs change to a gas through the process of sublimation.

The tendency of molecules in matter to move from an area of high concentration to low concentrations is **diffusion**. Putting a tea bag in water is an example of diffusion. The tea bag has a high concentration of tea: the water has a low concentration of tea. The tea molecules then have a tendency to move into the water until there is an equal concentration in each. The temperature of the water determines how fast the diffusion takes place. Heat is also being exchanged in this action. Some of the heat from the water is being transferred to the tea in the tea bag.

Heat (thermal energy) can be transferred. The movement of molecules in matter speeds up when heated. Heat can be transferred in three ways: conduction, convection, and radiation. **Conduction** is the transfer of energy through matter from molecule to molecule. For example, a spoon in a cup of hot soup becomes warmer because the heat from the soup is conducted along the spoon. **Convection** is the transfer of heat through a fluid (liquid or gas) caused by molecular motion. The motion causes currents. Convection is responsible for the rise and fall of macaroni in a pot of boiling water. In **radiation**, energy is transferred by molecules in electromagnetic waves. This energy comes in different wavelengths: radio, microwaves, infrared, visible light, ultraviolet light, x-rays, and gamma rays. The sun heats Earth through the process of radiation.

A **mixture** is made when two or more substances are physically combined but not chemically joined together. The substances still have the same properties as before they were combined and can be separated by ordinary physical means. Mixing table salt in water is an example of a mixture. The table salt dissolves in the water and seems to disappear. However, the table salt molecules actually take up the space between the water molecules. When the water is evaporated, the salt will remain. The table salt and water do not lose their own properties.

Name: ______________________ Date: ______________________

# Quick Check

## Matching

| | | | |
|---|---|---|---|
| e | 1. evaporation | a. | no gain or loss of matter during a change in form |
| a | 2. Law of Conservation of Mass | b. | the transfer of energy through matter from molecule to molecule |
| b | 3. conduction | c. | the transfer of heat through a fluid |
| c | 4. convection | d. | two or more substances are physically combined but not chemically joined together |
| d | 5. mixture | e. | the process of a liquid changing to a gas |

## Fill in the Blanks

6. The states of matter are solid, liquid, gas, and plasma.
7. Tearing paper changes its physical appearance (shape and size) but not its chemical makeup.
8. When a liquid is frozen, its molecules slow down.
9. Sublimation occurs when a solid changes directly to a gas without changing to a liquid first.
10. The tendency of molecules in matter to move from an area of high concentration to low concentrations is diffusion.

## Multiple Choice

11. Which of the following is a physical property of matter?
    (a.) color b. ability to rust c. flammability d. ability to tarnish

12. Molecules in a solid ________.
    a. are completely motionless b. stay in about the same position
    (c.) vibrate back and forth d. move around one another freely

13. How does water vapor in air form clouds?
    a. melting b. evaporation c. sublimation (d.) condensation

14. The change from a gas to a liquid is called ________.
    a. evaporation (b.) condensation c. melting d. sublimation

15. The temperature at which a substance changes from a liquid to a solid is the ________.
    a. boiling point (b.) freezing point c. changing point d. cooling point

Name: ______________________ Date: ______________

# Knowledge Builder

## Activity #1: Effects of Cooling and Heating Air

**Directions:** Stretch a large, empty balloon over the mouth of a flask. Place the flask on a stand over a Bunsen burner. Heat the flask. Observe the effect of heating on the balloon. Sketch the flask and balloon before and after adding heat.

| Before | After |
|---|---|
| | |

Conclusion: ______________________________________________

______________________________________________

______________________________________________

## Activity #2: Diffusion

**Directions:** Pour 250 mL of hot water in one cup. Measure and record the temperature: temperature ________°C. Pour 250 mL of cold water in the other cup. Measure and record the temperature: temperature ________°C. Place a tea bag in each cup and record the time: starting time ________. Record the temperature and your observations for the times in the data table below.

| Time (min.) | Cold (°C) | Observations | Hot (°C) | Observations |
|---|---|---|---|---|
| 1 min. | | | | |
| 2 min. | | | | |
| 3 min. | | | | |
| 4 min. | | | | |
| 5 min. | | | | |

Conclusion: ______________________________________________

______________________________________________

______________________________________________

Name: ______________________ Date: ______________

# Inquiry Investigation: Changing States of Matter

**Concept:**
- **Heat** (thermal energy) can be transferred.

**Purpose:** A question that asks what you want to learn from the investigation.

*Purpose:* Does an increase in temperature cause ice to change states?

**Hypothesis:** Write a sentence that predicts what you think will happen in the experiment. Your hypothesis should be clearly written. It should answer the question stated in the purpose.

*Hypothesis:* ______________________________________________

______________________________________________

**Procedure:** Carry out the investigation. This includes gathering the materials, following the step-by-step directions, and recording the data.

**Materials:**

hot plate
thermometer
stirring rod
ice cubes (100 mL)
watch with a second hand
250-mL beaker
water

**Experiment:**

Step 1: Pour 150 mL of water and 100 mL of ice into the beaker. Place the beaker on the hot plate.

Step 2: Put the thermometer into the ice/water mixture. After 30 seconds, record the temperature in the data table.

Step 3: Plug in the hot plate and turn the temperature knob to the medium setting. (**Warnings:** Do not touch the coils of the hot plate. Do not stir with the thermometer or allow it to rest on the bottom of the beaker while heating.)

Step 4: Every 30 seconds, read and record the temperature and physical state of the water until it begins to boil. Use the stirring rod to stir the contents of the beaker before making each temperature measurement.

**Results:**

Record the temperature and your observation for the times in the data table.

| Time (sec.) | Temperature (°C) | Physical State |
|---|---|---|
| | | |
| | | |
| | | |
| | | |
| | | |
| | | |
| | | |

Name: ______________________ Date: ______________

**Analysis:** Study the results of your experiment. Decide what the data means. This information can then be used to help you draw a conclusion about what you learned in your investigation.

Use your data to create a bar graph. Plot the temperature on the *y*-axis. Plot the time on the *x*-axis.

## Changing a Solid to a Liquid and a Gas

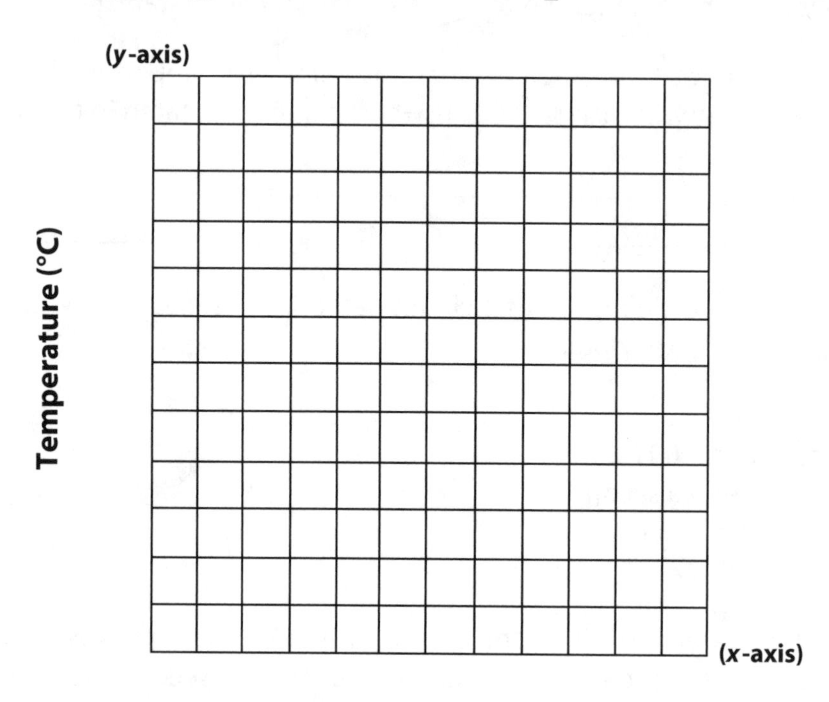


**Time (seconds)**

**Conclusion:** Write a brief description of what happened in the experiment and whether or not your hypothesis was correct.

_______________________________________________

_______________________________________________

_______________________________________________

_______________________________________________

_______________________________________________

_______________________________________________

_______________________________________________

_______________________________________________

_______________________________________________

# Unit 9: Chemical Changes in Matter
## Teacher Information

**Topic:** A chemical change happens at the atomic level.

**Standards:**
**NSES** Unifying Concepts and Processes, (A), (B)
**NCTM** Geometry, Number and Operations, Measurement, Data Analysis and Probability
See **National Standards** section (pages 64–65) for more information on each standard.

**Concepts:**
- A chemical change takes place when substances are combined and they change their chemical structure.
- The structures of each of the substances in a compound change at the atomic level.

**Naïve Concepts:**
- Chemical changes are perceived as additive, rather than interactive. After a chemical change, the original substances are perceived as remaining, even though they are altered.
- Formation of a new substance with new properties is a simple happening, rather than as the result of particle reengagement. (American institute of Physics, 2000)

**Science Process Skills:**
Students will be **measuring** pH levels of common substances. They will be **manipulating** materials to make **predictions** and **conducting an experiment** to determine the pH levels of common substances. Students will be **communicating** and **developing vocabulary** during the process of **collecting**, **recording**, **analyzing**, and **interpreting** data.

**Lesson Planner:**
1. Directed Reading: Introduce the concepts and essential vocabulary relating to the chemical changes of matter using the directed reading exercise on the Student Information pages.
2. Assessment: Evaluate student comprehension of the information in the directed reading exercise using the quiz located on the Quick Check page.
3. Concept Reinforcement: Strengthen student understanding of concepts with the activities found on the Knowledge Builder page.
4. Inquiry Investigation: Explore the pH levels of common substances. Divide the class into teams. Instruct each team to complete the Inquiry Investigation page.

**Extension:** The skin has an acidic pH, around 5.5. The frequent use of soap and detergents tends to decrease the pH of the skin and cause harm. Test the pH of liquid dish soaps to determine the kind least harmful to skin.

**Real World Application:** Milk that has gone sour in the refrigerator is due to a chemical reaction.

# Unit 9: Chemical Changes in Matter
## Student Information

A **chemical change** takes place when substances are combined and they change their chemical structure, producing a new substance. The new substance has properties different from the original substances. During a chemical reaction, atoms are rearranged into new substances. An animal digesting food is an example of a chemical change.

A chemical change happens at the atomic level. A **compound** is two or more substances that are chemically combined to form a new substance. The elements in a compound are held together by **chemical bonds**. Chemical bonds are formed between the atoms of the elements. There are two main types of chemical bonds.

- An **ionic bond** is formed when atoms transfer electrons. **Salt is an example of an ionic bond.** A sodium atom transfers an electron to a chlorine atom.

*Sodium ion Na* *Chloride ion Cl*

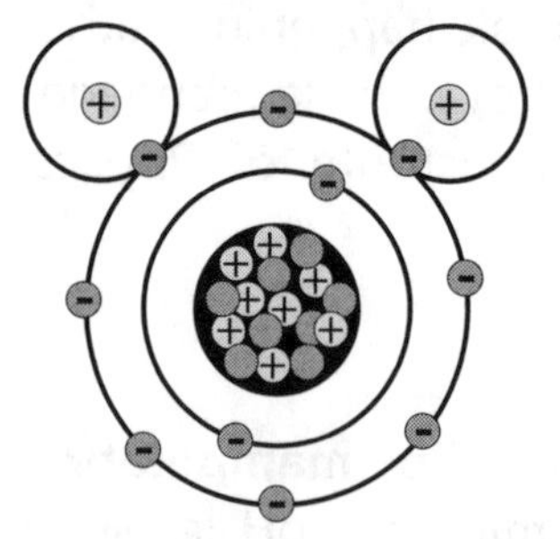

- A **covalent bond** is formed when atoms share electrons. **Water is an example of a covalent bond.** The hydrogen atoms share electrons with the oxygen atom.

The structures of each of the substances in a compound change at the atomic level. New substances form because the chemical bonds between atoms break. Then the atoms form new bonds between different atoms. The new substances are made up of a different combination of atoms and have different properties. This means they can only be separated by another chemical reaction. An example of a chemical compound is water. One atom of oxygen combines with two atoms of hydrogen to form a molecule of water. Water has the chemical formula $H_2O$. Both are invisible gases at room temperature before they are combined. After they are combined, they form water, a clear liquid at room temperature.

Chemical reactions are described using chemical equations. A **chemical equation** is a shorthand way to describe a chemical reaction between two or more substances. $2H_2 + O_2 \rightarrow 2H_2O$ is the chemical equation for water. The "+" sign means combine, and an arrow means "yield". $2H_2 + O_2 \rightarrow 2H_2O$ **means four atoms of hydrogen combine with two atoms of oxygen to yield two water molecules.** A **chemical formula** is a shorthand way to name the elements in a compound and their proportions. $H_2O$ is the chemical formula scientists use to represent the compound we call water.

During a chemical reaction, matter is not created or destroyed. All the atoms present at the start of the reaction are also present at the end of the reaction. This is known as the **Law of Conservation of Mass.**

Chemical compounds that taste sour, react with metal to produce hydrogen gas, and conduct electricity are called **acids**. Some foods, such as lemon juice, contain acid. However, many acids are too strong to taste or even touch. One of the strongest acids is sulfuric acid. Sulfuric acid is used in car batteries because it is a good conductor of electricity. Most acids are made of one or more hydrogen atoms. In the chemical formula for most acids, hydrogen is the first element. The chemical formula for hydrochloric acid is HCl.

Chemical compounds that taste bitter, feel slippery, dissolve oils and fats, and conduct electricity are called **bases**. Ammonia, lye, and bleach are bases. Like acids, strong bases can also be dangerous to taste or touch. Most bases are made of hydrogen and oxygen atoms linked together. These atoms are called hydroxides. The formula for hydroxide is OH. It is always written last in a chemical formula. The chemical name for lye is sodium hydroxide. The formula is NaOH.

Sometimes the substance we are testing is neither an acid nor a base. It is a neutral substance. When an acid and a base react chemically, they neutralize each other. The hydrogen from the acid combines with the base's hydroxide to form water. Water is a neutral substance. Salt is another neutral compound produced when acids and bases react chemically.

A **pH scale** is a tool used to determine the acidic or basic level of a substance. The pH scale ranges from 0 to 14. **Litmus paper** is an indicator used to tell if a substance is an acid or a base. Acids turn blue litmus paper red. Bases turn red litmus paper blue. The color of the paper matches up with the numbers on the pH scale to indicate what kind of substance is being tested. A substance with a pH of 7 is classified as a neutral (neither acid nor base). A substance with a pH below 7 is classified as an acid. A substance with a pH above 7 is classified as a base.

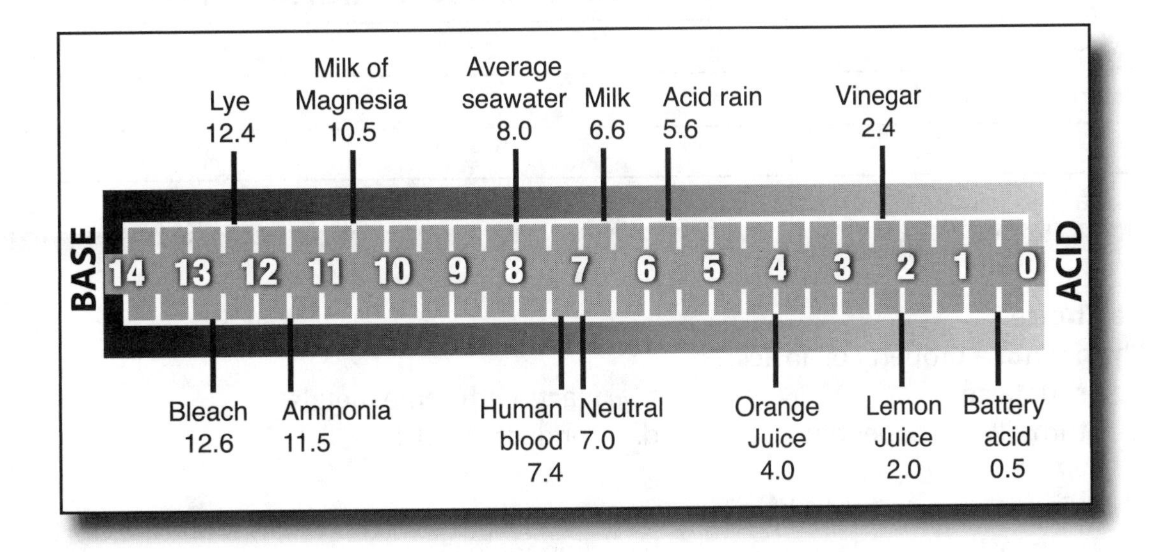

Name: ______________________________ Date: ______________

# Quick Check

## Matching

| | | | |
|---|---|---|---|
| ______ 1. | chemical change | a. | atoms share electrons |
| ______ 2. | ionic bond | b. | a new substance is formed |
| ______ 3. | covalent bond | c. | tastes sour |
| ______ 4. | acid | d. | atoms transfer electrons |
| ______ 5. | base | e. | tastes bitter |

## Fill in the Blanks

6. During a ______________ ______________, atoms are rearranged into new substances.
7. The elements in a compound are held together by ______________ ______________.
8. A ______________ ______________ is a shorthand way to name the elements in a compound and their proportions.
9. A ______________ ______________ is a tool used to determine the acidic or basic level of a substance.
10. ______________ ______________ is an indicator used to tell if a substance is an acid or a base.

## Data Table

Complete the data table by classifying each example as a chemical equation or chemical formula.

| Example | Chemical Equation/Chemical Formula |
|---|---|
| 11. $CO_2$ | |
| 12. $2H_2 + O_2 \rightarrow 2H_2O$ | |
| 13. $Fe + S \rightarrow FeS$ | |
| 14. NaCl | |

## Multiple Choice

15. Which is **not** a property of an acid?
    a. sour taste
    b. reacts with some metals
    c. turns litmus paper blue
    d. conducts electricity

16. Which is **not** a property of a base?
    a. sour taste
    b. slippery feel
    c. turns litmus paper blue
    d. dissolve oils and fats

Name: ______________________ Date: ______________________

# Knowledge Builder

## Activity #1: pH Scale

**Directions:** Using the pH Scale, complete the data table. Identify each substance as acid, base, or neutral. Bleach has been done for you.

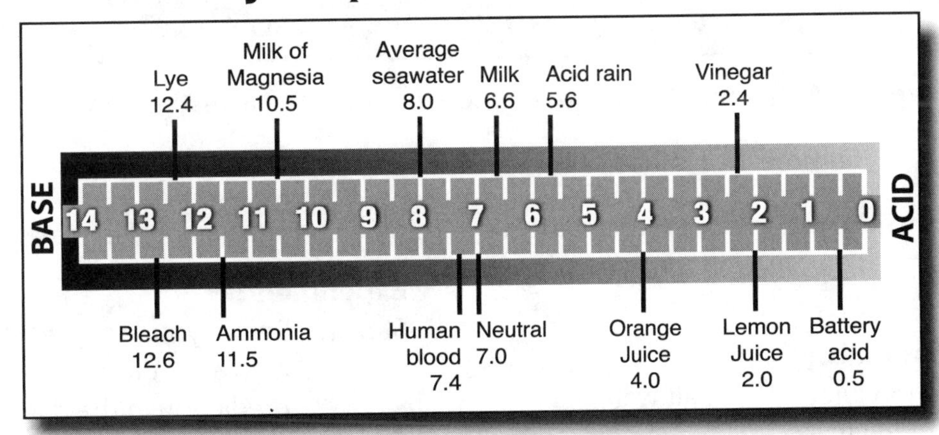


| pH Values of Common Substances Table | | | | |
|---|---|---|---|---|
| **Substance** | **pH Level** | **Acid** | **Base (alkaline)** | **Neutral** |
| 1. lemon juice | 2.0 | | | |
| 2. milk | 6.6 | | | |
| 3. bleach | 12.6 | | X | |
| 4. battery acid | 0.5 | | | |
| 5. orange juice | 4.0 | | | |
| 6. ammonia | 11.5 | | | |
| 7. vinegar | 2.4 | | | |
| 8. lye | 12.4 | | | |
| 9. blood | 7.4 | | | |

## Activity #2: Chemical Formulas

**Directions:** Research the compounds below. Complete the data table by writing the formula for each compound.

| **Compound** | **Formula** |
|---|---|
| 1. sodium hydroxide | |
| 2. carbon dioxide | |
| 3. sodium chloride | |
| 4. calcium carbonate | |
| 5. potassium nitrate | |

Name: ______________________________ Date: ____________________

# Inquiry Investigation: Acids and Bases

**Concept:**
- The **pH scale** is a tool used to determine the acidic or basic level of a substance.

**Purpose:** A statement telling the reason for doing the investigation.

*Purpose:* Use litmus paper to determine the pH level of common substances.

**Procedure:** Carry out the investigation. This includes gathering the materials, following the step-by-step directions, and recording the data.

**Materials:**

lemon juice
bottled water
orange juice
vinegar
tap water
salt water
lemon-lime soda
ammonia
baking soda solution
9 plastic cups
red and blue litmus paper
pH color chart

**Experiment:**

Step 1: Before you test, predict whether the substance is an acid, base, or neutral. Record your prediction in the date table below.

Step 2: Place a small amount of the solution in a cup. Label the cup with the name of the substance.

Step 3: Carefully dip a strip of red and blue litmus paper into the substance and compare the color of the papers to the pH color chart.

Step 4: Record the pH value and whether the substance is an acid, base, or neutral.

Step 5: Repeat steps 1–4 for each of the substances.

**Results:**
Complete the data table.

| Substance | Prediction Acid/Base/Neutral | Paper Color | pH Level | Acid/Base/Neutral |
|---|---|---|---|---|
| 1. lemon juice | | | | |
| 2. salt water | | | | |
| 3. bottled water | | | | |
| 4. lemon-lime soda | | | | |
| 5. orange juice | | | | |
| 6. ammonia | | | | |
| 7. vinegar | | | | |
| 8. baking soda solution | | | | |
| 9. tap water | | | | |

**Conclusion:**

1. Which substance is the strongest acid? ____________________
2. Which substance is the strongest base? ____________________
3. What would happen to the pH of vinegar if ammonia were added to it? ____________________

____________________________________________

# Unit 10: Evidence of Chemical Changes
## Teacher Information

**Topic:** During a chemical reaction, a new substance is formed.

**Standards:**
**NSES** Unifying Concepts and Processes, (A), (B)
**NCTM** Measurement and Data Analysis
See **National Standards** section (pages 64–65) for more information on each standard.

**Concepts:**
- Substances react chemically in characteristic ways with other substances to form new substances with different properties.

**Naïve Concepts:**
- Chemical changes are perceived as additive, rather than interactive. After a chemical change, the original substances are perceived as remaining, even though they are altered.

**Science Process Skills:**
Students will be **measuring** volume. They will be **predicting** and **observing** what happens during a chemical reaction. They will be **inferring** why the temperature changes when the substances are mixed. Students will be **communicating and developing vocabulary** relating to chemical reactions.

**Lesson Planner:**
1. Directed Reading: Introduce the concepts and essential vocabulary relating to the chemical changes of matter using the directed reading exercise on the Student Information page.
2. Assessment: Evaluate student comprehension of the information in the directed reading exercise using the quiz located on the Quick Check page.
3. Concept Reinforcement: Strengthen student understanding of concepts with the activities found on the Knowledge Builder page. **Materials Needed:** Activity #2—three different brands of microwave popcorn, access to a microwave, bowls to sort and count kernels
4. Inquiry Investigation: Explore what takes place during chemical reactions. Divide the class into teams. Instruct each team to complete the Inquiry Investigation page. (**Warning:** Due to the materials used, this activity is an upper-level activity and should be monitored by an adult at all times! Goggles are recommended. Do not mix more than the recommended amounts. Do the activity in a well ventilated area. Use plastic gloves when handling the materials. Be sure to add all ingredients in the proportions given. Before disposing of the liquid waste material, dilute it with a lot of water to stop the chemical reaction.)

**Extension:** Baking cookies, cakes, and breads are examples of chemical changes. The gases given off during the chemical reactions taking place make them rise as they bake.

**Real World Application:** Iron tools left outside in the rain for an extended period of time will eventually rust.

# Unit 10: Evidence of Chemical Changes

Student Information

Matter can change physically. Substances boil, melt, freeze, and mix. But not all changes in matter are physical ones. In some changes, the properties of the substances change. Then new substances are formed. The kind of change that creates a new substance is called chemical change. In a **chemical change**, a chemical reaction takes place. When a scientist combines hydrogen and oxygen to make water, a chemical reaction is taking place. Hydrogen and oxygen are two elements that have their own properties. Hydrogen and oxygen are gases that do not have exact size and shape. They combine, or react, to form water, a new substance. The new substance has a new set of properties. Water is a clear liquid that has size but does not have a definite shape.

### Evidence of a Chemical Reaction

1. the formation of a gas
2. an odor is given off
3. a change in color
4. the production of heat or light
5. the formation of a solid

Chemical reactions involve two main kinds of changes that can be observed—the formation of new substances and changes in energy.

### New Substance Formed

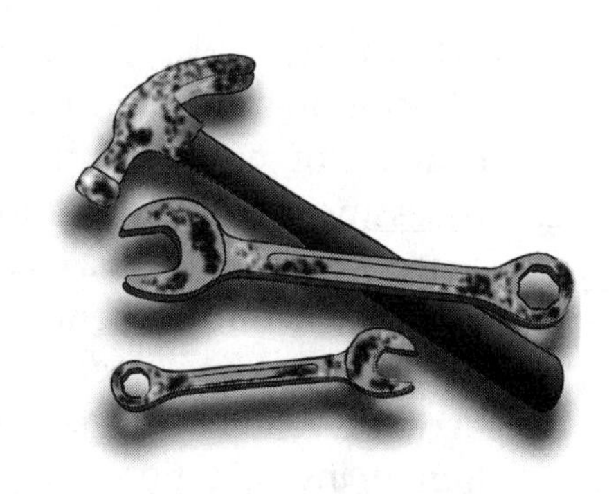

- **Oxidation:** Oxidation is the combining of oxygen with other substances. Fire or burning is a chemical reaction. It is the process of the fuel combining rapidly with oxygen; this process is called rapid oxidation. When plant matter, such as grass clippings, decays, the plant material is slowly combining with oxygen. This is a process called slow oxidation. Rust is also an example of slow oxidation. Heat is given off in all types of oxidation.
- **Precipitation:** Precipitation is the formation of a solid in a solution during a chemical reaction. When the reaction occurs, the precipitate settles out as a solid in the bottom of the container of liquid.

### Changes in Energy

- **Endothermic Reaction:** In an endothermic reaction, energy is absorbed. Frying an egg is an example of an endothermic reaction. You must keep adding heat, or the reaction will stop.
- **Exothermic Reaction:** In an exothermic reaction, energy is given off in the form of heat. Burning wood is an exothermic reaction because heat is given off.

Name: ______________________________ Date: ____________________

# Quick Check

### Matching

| | | | |
|---|---|---|---|
| _____ 1. | chemical change | a. | energy is absorbed |
| _____ 2. | oxidation | b. | a new substance is formed |
| _____ 3. | precipitation | c. | energy is given off |
| _____ 4. | endothermic reaction | d. | combining of oxygen with other substances |
| _____ 5. | exothermic reaction | e. | formation of a solid in a solution during a chemical reaction |

### Fill in the Blanks

6. When hydrogen and oxygen combine, ______________ is formed.
7. Rust is an example of slow ______________.
8. ______________ is given off in all types of oxidation.
9. Frying an egg is an example of an ______________ reaction.
10. Burning wood is an ______________ reaction because heat is given off.

### Data Table

Complete the data table by classifying each example as a chemical or physical change.

| Change | Type of Change |
|---|---|
| 11. freezing water | |
| 12. rotting wood | |
| 13. cake baking | |
| 14. burning paper | |
| 15. bread molding | |
| 16. iron rusting | |
| 17. crushing rock | |
| 18. drying clothes | |

### Identify

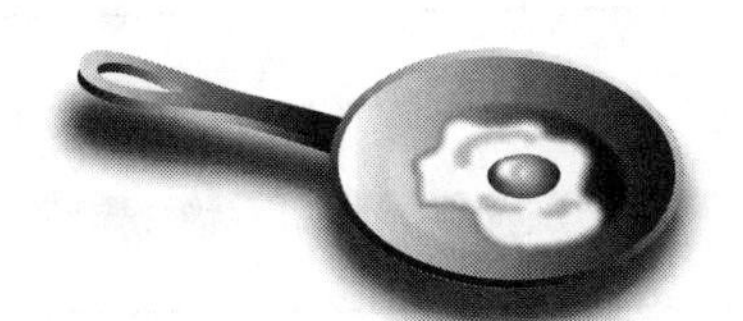

19. What kind of reaction occurs when an egg is cooking?

______________________

20. What kind of reaction occurs when a candle is burning?

______________________

Name: ______________________________ Date: ______________________

# Knowledge Builder

## Activity #1: Endothermic and Exothermic Reactions

**Directions:** Complete the data table about changes in energy in chemical reactions. Classify each change as endothermic or exothermic, and indicate if energy is released or absorbed.

| Example | Type of Reaction | Released/Absorbed |
|---|---|---|
| 1. rotting wood | | |
| 2. rusting steel | | |
| 3. photosynthesis | | |
| 4. cooking an egg | | |
| 5. heating with coal | | |

## Activity #2: Physical Changes in Matter

**Directions:** When popcorn pops, an amazing change takes place inside the kernel. The solid material in the kernel becomes a liquid and then a gas. The gas molecules move farther apart as they get hotter and hotter. Eventually, the molecules can no longer be contained inside the kernel, and they start popping out of the shells. Complete the following activity, and then decide if corn popping is an example of a physical or chemical change. Select three brands of microwave popcorn. Pop one bag of each brand (follow package directions). After a bag has popped, count all the popped and unpopped kernels. Record the total in the Number of Kernels at Start column in the data table. Now separate the popped and unpopped kernels into different piles. Count the number of popped kernels. Record the answer. Count the number of unpopped kernels. Record the answer. The number of kernels popped can be written as a percentage. To find the percentage: (1) divide the number of popped kernels by the total number of kernels you had at the start, and (2) multiply your answer by 100. Your new answer is the percent of popped kernels. Record the answer.

| Brand | Number of Kernels at Start | Number of Popped Kernels | Number of Unpopped Kernels | Percentage of Popped Kernels (%) |
|---|---|---|---|---|
| | | | | |
| | | | | |
| | | | | |

1. Which brand of popcorn popped the most kernels? ______________________
2. Based on your investigation and observations, is corn popping a chemical or physical change? Explain your answer. ______________________

______________________

______________________

Name: ____________________ Date: ____________________

# Inquiry Investigation: Oxidation

**Concept:**
- **Oxidation** is a chemical reaction involving oxygen.

**Purpose:** A question that asks what you want to learn from the investigation.

*Purpose:* Does oxidation affect the properties of metal?

**Hypothesis:** Write a sentence that predicts what you think will happen in the experiment. Your hypothesis should be clearly written. It should answer the question stated in the purpose.

*Hypothesis:* ______________________________
______________________________

**Procedure:** Carry out the investigation. This includes gathering the materials, following the step-by-step directions, and recording the data.

**Materials:**
one small ball of steel wool
magnet
measuring spoons (5 mL = 1 tsp.)
matches
water (approx. 125 mL)
paper towels or coffee filter
20 mL bleach
candle
plastic gloves
10 mL vinegar
2 small jars
metal spoon
goggles

**Experiment:**
**Part I (Adult supervision is required for this experiment.)**

Step 1: Pull off a small amount of steel wool. Examine the steel wool. Record your observations in the data table. Bring a magnet up to the steel wool. Is the steel wool attracted to the magnet? Record your observation in the data table.

Step 2: Make a small ball of the steel wool and place it in a jar. Put enough water in the jar to cover the ball (approx. 125 mL). Let the steel wool sit in the water for five minutes. Record your observations in the data table.

Step 3: Add 20 mL of bleach and 10 mL of vinegar to the water. Record your observations in the data table.

Step 4: Let the jar sit for 5 minutes. Record your observations in the data table.

**Results:**

| Times Observed | Steel Wool Characteristics |
|---|---|
| Before Experiment | |
| In Water | |
| In Bleach/Vinegar Solution | |

Name: ______________________ Date: ______________

**Part II**

Step 1: Remove the steel wool from the jar. Allow the liquid to sit until the mixture settles and an orange powder forms at the bottom of the jar.

Step 2: Drain off the liquid carefully so you do not spill the orange powder out of the jar. (Dilute the bleach/vinegar/water mixture with plenty of water before disposing of it to stop the reaction.)

Step 3: Fill the jar with the powder with water and stir it.

Step 4: Allow the orange substance to settle to the bottom again. Once it has settled out again, drain off most of the water.

Step 5: Place the coffee filter in the other jar. Pour the remaining liquid and powder into the filter.

Step 6: Spread out the filter with the orange powder and allow it to dry.

Step 7: Closely examine the orange powder. Test the orange substance to see if it is attracted to a magnet. Record your observations in the data table.

**Part III**

Step 1: Place the orange substance in an old spoon. Hold the spoon over a candle flame. Observe the powder as it is heated. Record your observations in the data table.

Step 2: After the orange powder has changed color completely, put the substance on the coffee filter again. Is the steel powder attracted to the magnet? Record your observations in the data table.

**Results:**

| Times Observed | Orange Powder Characteristics |
|---|---|
| In Water | |
| After Drying | |
| After Heating | |

**Analysis:**

Study the results of your experiment. Decide what the data means.

1. Why do you think the steel wool turned orange and left a powder in the jar? ______________________

2. Why did the orange powder change colors when it was heated? ______________________

3. Why wasn't the orange powder attracted to the magnet? ______________________

**Conclusion:** Write a brief description of what happened in the experiment and whether or not your hypothesis was correct.

______________________

Name: ____________________ Date: ____________

# Inquiry Investigation Rubric

| Category | 4 | 3 | 2 | 1 |
|---|---|---|---|---|
| Participation | Used time well, cooperative, shared responsibilities, and focused on the task. | Participated, stayed focused on task most of the time. | Participated, but did not appear very interested. Focus was lost on several occasions. | Participation was minimal OR student was unable to focus on the task. |
| Components of Investigation | All required elements of the investigation were correctly completed and turned in on time. | All required elements were completed and turned in on time. | One required element was missing/or not completed correctly. | The work was turned in late and/or several required elements were missing and/or completed incorrectly. |
| Procedure | Steps listed in the procedure were accurately followed. | Steps listed in the procedure were followed. | Steps in the procedure were followed with some difficulty. | Unable to follow the steps in the procedure without assistance. |
| Mechanics | Flawless spelling, punctuation, and capitalization. | Few errors. | Careless or distracting errors. | Many errors. |

**Comments:**

# National Standards Related to Chemistry

## National Science Education Standards (NSES) Content Standards (NRC, 1996)

National Research Council (1996). *National Science Education Standards.* Washington, D.C.: National Academy Press.

**Unifying Concepts K–12**

Systems, order, and organization
Evidence, models, and explanation
Change, constancy, and measurement
Form and function

**NSES Content Standard A: Inquiry**

- Abilities necessary to do scientific inquiry
- Understanding about inquiry

**NSES Content Standard B: 5–8 Properties and Changes of Properties in Matter**

- A substance has characteristic properties—density, boiling point, solubility. A mixture can often be separated into original substances.
- Substances react chemically in characteristic ways with other substances to form new substances with different properties; conservation of mass in chemical reactions; substances are grouped if they react in similar ways.
- Chemical elements do not break down by ordinary chemical laboratory treatments, such as heating, exposure to electric current, or reaction with acids. There are over 100 known elements that combine to form compounds (NRC, 1996).

**NSES Content Standard B: 5–8 Transfer of Energy**

- Energy is a property of many substances and is associated with heat, light, electricity, mechanical motion, sound, nuclei, and the nature of a chemical; energy is transferred in many ways.
- Electrical circuits provide a means of transferring electrical energy when heat, light, sound, and chemical changes are produced.
- In most chemical and nuclear reactions, energy is transferred into or out of the system; heat, light, mechanical motion, or electricity might be involved (NRC, 1996).

**NSES Content Standard C: Life Science 5–8**

- Structure and function in living systems

**NSES Content Standard D: Earth and Space**

- Structure of the earth system

**NSES Content Standard E: Science and Technology 5–8**

- Abilities of technological design
- Understanding about science and technology

**NSES Content Standard F: Science in Personal and Social Perspectives 5–8**

- Personal health
- Science and technology in society

**NSES Content Standard G: History and Nature of Science 5–8**

- Science as a human endeavor
- Nature of science
- History of science

# National Standards Related to Chemistry (cont.)

## Principles and Standards For School Mathematics (NCTM, 2000)

National Council for Teachers of Mathematics. (2000). *Principles and Standards for School Mathematics.* Reston,VA: National Council for Teachers of Mathematics.

**Number and Operations**
- Understand numbers, ways of representing numbers, relationships among numbers, and number systems.
- Understand meanings of operations and how they relate to one another.
- Compute fluently and make reasonable estimates.

**Algebra**
- Understand patterns, relations, and functions.
- Represent and analyze mathematical situations and structures using algebraic symbols.
- Use mathematical models to represent and understand quantitative relationships.
- Analyze change in various contexts.

**Geometry**
- Analyze characteristics and properties of two- and three-dimensional geometric shapes and develop mathematical arguments about geometric relationships.
- Specify locations and describe spatial relationships using coordinate geometry and other representational systems.
- Apply transformations and use symmetry to analyze mathematical situations.
- Use visualization, spatial reasoning, and geometric modeling to solve problems.

**Measurement**
- Understand measurable attributes of objects and the units, systems, and processes of measurement.
- Apply appropriate techniques, tools, and formulas to determine measurements.

**Data Analysis and Probability**
- Formulate questions that can be addressed with data, and collect, organize, and display relevant data to answer them.
- Select and use appropriate statistical methods to analyze data.
- Develop and evaluate inferences and predictions that are based on data.
- Understand and apply basic concepts of probability.

# Science Process Skills: Chemistry

**Introduction:** Science is organized curiosity, and an important part of this organization is the thinking skills or information-processing skills. We ask the question "Why?" and then must plan a strategy for answering the question or questions. In the process of answering our questions, we make and carefully record observations, make predictions, identify and control variables, measure, make inferences, and communicate our findings. Additional skills may be called upon depending on the nature of our questions. In this way, science is a verb involving active manipulation of materials and careful thinking. Science is dependent upon language, math, and reading skills as well as the specialized thinking skills associated with identifying and solving problems.

**BASIC PROCESS SKILLS**

**Classifying:** Grouping, ordering, arranging, or distributing objects, events, or information into categories based on properties or criteria, according to some method or system.

> Example – The skill is being demonstrated if the student is …
> Grouping substances by their physical properties into categories. These categories might include states of matter, such as solids, liquids, and gases; or elements, compounds, or mixtures.

# Science Process Skills: Chemistry (cont)

**Observing:** Using the senses (or extensions of the senses) to gather information about an object or event.

> Example – The skill is being demonstrated if the student is …
> Seeing and describing the physical or chemical properties of a substance.

**Measuring:** Using both standard and nonstandard measures or estimates to describe the dimensions of an object or event. Making quantitative observations.

> Example – The skill is being demonstrated if the student is …
> Using a graduated cylinder to measure the volume of a liquid or an irregularly shaped object or using a ruler to measure the volume of a block.

**Inferring:** Making an interpretation or conclusion based on reasoning to explain an observation.

> Example – The skill is being demonstrated if the student is…
> Stating that a chemical change has taken place by observations or tests conducted on the new substance.

**Communicating:** Communicating ideas through speaking or writing. Students may share the results of investigations, collaborate on solving problems, and gather and interpret data both orally and in writing. Uses graphs, charts, and diagrams to describe data.

> Example – The skill is being demonstrated if the student is …
> Describing an event or a set of observations. Participating in brainstorming and hypothesizing before an investigation. Formulating initial and follow-up questions in the study of a topic. Summarizing data, interpreting findings, and offering conclusions. Questioning or refuting previous findings. Making decisions or using a graph to show the relationship between temperature and the decrease in size of a melting ice cube over time.

**Predicting:** Making a forecast of future events or conditions in the context of previous observations and experiences.

> Example – The skill is being demonstrated if the student is …
> Stating how fast an ice cube will melt at a specific temperature based on data collected through experimentation and observation.

**Manipulating Materials:** Handling or treating materials and equipment skillfully and effectively.

> Example – The skill is being demonstrated if the student is …
> Arranging equipment and materials needed to conduct an investigation and setting up and conducting an experiment to determine air is matter by showing that it has mass and takes up space.

**Replicating:** Performing acts that duplicate demonstrated symbols, patterns, or procedures.

> Example – The skill is being demonstrated if the student is …
> Using a double-pan balance following procedures previously demonstrated or modeled by another person or following a procedure to set up a double-pan balance made from common materials.

# Science Process Skills: Chemistry (cont.)

**Using Numbers:** Applying mathematical rules or formulas to calculate quantities or determine relationships from basic measurements.

> Example – The skill is being demonstrated if the student is …
> Computing the density of a substance using the formula **Density = Mass divided by Volume**.

**Developing Vocabulary:** Specialized terminology and unique uses of common words in relation to a given topic need to be identified and given meaning.

> Example – The skill is being demonstrated if the student is …
> Using context clues, working definitions, glossaries or dictionaries, word structure (roots, prefixes, suffixes), and synonyms and antonyms to clarify meaning.

**Questioning:** Questions serve to focus inquiry, determine prior knowledge, and establish purposes or expectations for an investigation. An active search for information is promoted when questions are used.

> Example – The skill is being demonstrated if the student is …
> Using what is already known about a topic or concept to formulate questions for further investigation; hypothesizing and predicting prior to gathering data; or formulating questions as new information is acquired.

**Using Cues:** Key words and symbols convey significant meaning in messages. Organizational patterns facilitate comprehension of major ideas. Graphic features clarify textual information.

> Example – The skill is being demonstrated if the student is …
> Listing or underlining words and phrases that carry the most important details or relating key words together to express a main idea or concept.

## INTEGRATED PROCESS SKILLS

**Creating Models:** Displaying information by means of graphic illustrations or other multisensory representations.

> Example – The skill is being demonstrated if the student is …
> Drawing a graph or diagram; constructing a three-dimensional object, using a digital camera to record an event, constructing a chart or table, or producing a picture or diagram that illustrates information about the molecular structures of the states of matter.

**Formulating Hypotheses:** Stating or constructing a statement that is testable about what is thought to be the expected outcome of an experiment (based on reasoning).

> Example – The skill is being demonstrated if the student is …
> Making a statement to be used as the basis for an experiment, i.e., If heat is added or taken away, matter changes from one state to another.

**Generalizing:** Drawing general conclusions from particulars.

> Example – The skill is being demonstrated if the student is …
> Making a summary statement following analysis of experimental results, i.e., Matter changes from a solid to a liquid by adding heat.

# Science Process Skills: Chemistry (cont.)

**Identifying and Controlling Variables:** Recognizing the characteristics of objects or factors in events that are constant or change under different conditions and that can affect an experimental outcome, keeping most variables constant while manipulating only one variable.

> Example – The skill is being demonstrated if the student is …
> Listing or describing the factors that would influence the outcome of an experiment, such as the temperature and humidity of the air during an evaporation experiment.

**Defining Operationally:** Stating how to measure a variable in an experiment, defining a variable according to the actions or operations to be performed on or with it.

> Example – The skill is being demonstrated if the student is …
> Defining such things as heat in the context of materials and actions for a specific activity. Heat energy gain or loss can be measured by the increase or decrease in temperature.

**Recording and Interpreting Data:** Collecting bits of information about objects and events, which illustrate a specific situation; organizing and analyzing data that has been obtained; and drawing conclusions from it by determining apparent patterns or relationships in the data.

> Example – The skill is being demonstrated if the student is …
> Recording data (taking notes, making lists/outlines, recording numbers on charts/graphs, making tape recordings, taking photographs, writing numbers of results of observations/measurements) from the series of experiments to determine if air is matter and forming a conclusion that relates trends in data to variables.

**Making Decisions:** Identifying alternatives and choosing a course of action from among alternatives after basing the judgment for the selection on justifiable reasons.

> Example – The skill is being demonstrated if the student is …
> Identifying alternative ways to solve a problem through the use of physical or chemical properties of a substance; analyzing the consequences of each alternative, such as cost, the effect on other people or the environment; using justifiable reasons as the basis for making choices; and choosing freely from the alternatives.

**Experimenting:** Being able to conduct an experiment, including asking an appropriate question, stating a hypothesis, identifying and controlling variables, operationally defining those variables, designing a "fair" experiment, and interpreting the results of an experiment.

> Example – The skill is being demonstrated if the student is …
> Utilizing the entire process of designing, building, and testing various substances to solve a problem. Arranging equipment and materials to conduct an investigation, manipulating the equipment and materials, and conducting the investigation. An experiment was designed and conducted in the "Is Air Matter?" activity.

# Definitions of Terms

**Acids** are compounds that produce hydrogen ions in solutions.

**Adding or taking away heat** changes matter from one state to another.

**Adhesion** is the attraction between unlike substances.

**Alchemy** is one of the earliest forms of chemistry.

**Analytical chemistry** deals with the composition of substances.

**Archimedes' Principle** states that an object placed in a liquid seems to lose an amount of weight equal to the amount of fluid it displaces.

The **atomic mass numbers** of elements are often used in place of atomic weights. The atomic mass is the number of protons and neutrons in an atom of the element.

**Atomic symbols** are standard chemical abbreviations used to indicate a specific element.

**Atomic Theory** states that matter is made up of atoms.

**Atomic weights** are determined by comparing the element with an atom of carbon 12, which is assigned the weight of 12 units.

**Atoms** are the smallest part of an element.

**Bases** are compounds that produce hydroxide ions in solutions.

**Bernoulli's Principle** reveals that the pressure in a moving stream of fluid is less than that in the surrounding fluid.

**Biochemistry** is the study of the chemistry of living organisms.

**Capillarity** is the spontaneous movement of liquids up or down narrow tubes.

**Carbon dioxide** is a colorless, odorless, incombustible gas.

**Changes in matter** can be physical or chemical changes.

**Chemical bonds** are formed between the atoms of the elements.

When a **chemical change** takes place, a new substance with new properties is formed.

A **chemical equation** is a shorthand way to describe a chemical reaction between two or more substances.

A **chemical formula** is a shorthand way to name the elements in a compound and their proportions.

**Chemical properties** of matter are related to the atoms and molecules of the substance.

# Definitions of Terms (cont.)

**Chemistry** is the study of the structure, properties, and composition of substances and the changes to the substances. (Hewitt, 2002)

**Cohesion** is the attraction of like substances.

A **colloid** is formed when larger particles of matter are suspended in a solid, liquid, or gas.

**Compounds** are formed during a chemical change.

**Condensation** occurs when a substance turns from a gas to a liquid.

**Conduction** occurs when heat is transferred by collisions of the particles of the substance.

**Convection** occurs when heat is transferred into gas or liquid by the currents in the heated fluid.

A **covalent bond** is formed when atoms share electrons.

**Density** is the mass per unit volume.

**Diffusion** happens when a substance moves from an area of higher concentration to an area of lower concentration.

**Distillation** is a process by which a liquid is turned into a vapor and condensed back into a liquid.

**Elasticity** means that a solid can be bent or twisted a limited amount, and it will return to its former shape.

**Electrons** are the particles in atoms that have a negative charge.

**Elements** are substances made up of only one kind of atom that cannot be divided by ordinary chemical means.

In an **endothermic reaction**, energy is absorbed.

**Evaporation** occurs when a substance is changed from a liquid to a gas.

In an **exothermic reaction**, energy is given off in the form of heat.

**Families** are groups of elements with similar physical and chemical properties.

**Fermentation** is the production of alcohol from sugar through the action of yeast or bacteria.

**Filtration** uses porous materials to separate solids from liquids.

**Fluids** are any material that doesn't maintain a definite shape.

**Freezing** is the change from a liquid state to a solid state.

**Freezing point** is the temperature at which a substance changes from a liquid to a solid.

# Definitions of Terms (cont.)

**Gases** have no definite volume or shape.

An **iatrochemist** is another name for an alchemist in the 1500s and 1600s.

**Inorganic chemistry** is the study of compounds without carbon.

An **ionic bond** is formed when atoms transfer electrons.

**Isotopes** are elements with different numbers of neutrons.

The **Law of Buoyancy** states that an object placed into a liquid seems to lose an amount of weight equal to the amount of fluid it displaces.

The **Law of Combining Volumes** states that elements combine in definite proportions by volume to form compounds.

The **Law of Conservation of Matter** states that matter cannot be created or destroyed.

**Liquids** have a definite volume but no definite shape.

**Litmus paper** is an indicator used to tell if a substance is an acid or a base.

**Mass** is the amount of matter present.

**Matter** is anything that takes up space and has mass.

**Melting** is the change from a solid state to a liquid state.

**Melting point** is the temperature at which a substance changes from a solid to a liquid.

**Mixtures** are formed when two or more substances are mixed but are not chemically combined. They can be separated by ordinary physical means.

**Molecular Theory** states that matter is made up of molecules.

**Molecules** are the smallest part of a compound that still has the properties of the compound.

**Organic chemistry** is the study of substances containing carbon.

**Neutrons** are the particles in atoms that are neutral or have no charge.

**Oxidation** is the combining of oxygen with other substances.

**Oxygen** is a colorless, odorless, and tasteless gas.

The **Periodic Law** states that an element's properties depend upon its atomic weight.

The **Periodic Table of Elements** is a tabular method of displaying the chemical elements.

**Periods** are the horizontal rows in the Periodic Table of the Elements.

# Definitions of Terms (cont.)

A **pH scale** is a tool used to determine the acidic or basic level of a substance.

**Physical changes** in matter change the form or appearance of matter but not its chemical makeup. This may include adding or taking away heat energy to change states.

**Physical chemistry** deals with the study of heat, electricity, and other forms of energy in chemical processes.

**Physical properties** of matter include mass, volume, and density.

**Plasma** has no definite shape or volume and is a highly energized gas.

The **principle of conservation of mass** states that all atoms present at the start of the reaction are also present at the end of the reaction.

**Precipitation** is the formation of a solid in a solution during a chemical reaction.

**Protons** are the particles in atoms that have a positive charge.

A **quantum of energy** is the amount of energy needed to move an electron from its current level to the next higher level.

**Quantum leap** describes an abrupt change.

In the **quantum mechanical model**, electrons have a restricted value, but they do not have a specified path around the nucleus.

**Radiation** is energy transferred by electromagnetic waves.

**Salts** are formed when acids and bases are combined.

**Solids** have a definite shape and volume.

**Solutions** are formed when substances dissolve in a solid or liquid.

The four **states of matter** are solid, liquid, gas, and plasma.

**Sublimation** is when a substance changes directly from a solid to a gas without first changing to a liquid. An example of this is mothballs.

**Surface tension** is a property of liquids arising from unbalanced molecular cohesive forces at or near the surface.

A **thermoscope** was the first thermometer developed by Galileo.

**Viscosity** is a liquid's resistance to flow.

**Volume** is the amount of space something takes up.

# Answer Keys

***Historical Perspective***

**Quick Check (page 7)**

*Matching*

1. b 2. c 3. e 4. d 5. a

*Fill in the Blanks*

6. Atomic Theory 7. Organic chemistry
8. carbon 9. oxygen 10. atom, physics

*Time Line*

11. chemistry 12. 50 13. Jean Beguin
14. *Sceptical Chymist* 15. Joseph Priestly

***Chemistry***

**Quick Check (page 12)**

*Matching*

1. b 2. a 3. e 4. d 5. c

*Fill in the Blanks*

6. solids, liquids 7. Distillation 8. solid, gas
9. Physical chemistry 10. interact

*Multiple Choice*

11. d 12. b 13. c 14. a

***Elements***

**Quick Check (page 18)**

*Matching*

1. e 2. c 3. b 4. d 5. a

*Fill in the Blanks*

6. James Chadwick 7. atomic, mass, numbers
8. electrons 9. element 10. 118

*Multiple Choice*

11. b 12. a

*Reading a Table*

13. Iron 14. K 15. Calcium
16. Au 17. Boron

***The Four States of Matter***

**Quick Check (page 22)**

*Matching*

1. c 2. d 3. e 4. b 5. a

*Fill in the Blanks*

6. volume, mass
7. color, shape, smell, taste, texture
8. solid, liquid, gas, plasma
9. Plasma 10. gas

*Data Table*

11. ice 12. fish 13. water
14. milk 15. air 16. exhaust

*Venn Diagram*

Solid-definite shape

Both-made of matter, have volume and mass, made up of tiny particles in constant motion

Liquid-no definite shape

**Inquiry Investigation: Air (page 24)**

*Hypothesis:* Air has weight, and it can be measured.

*Observation:* The stick was evenly balanced. When one balloon was popped, the stick was unbalanced.

*Conclusion:* When one balloon was popped, the stick was unbalanced. The end tied to the popped balloon rose and the end tied to the inflated balloon fell. This indicates the air in the balloon has weight. The hypothesis that air has weight is correct.

***The Structure of Matter***

**Quick Check (page 27)**

*Matching*

1. a 2. c 3. b 4. e 5. d

*Fill in the Blanks*

6. atoms 7. Molecular Theory 8. mass
9. solid, liquid, gas, plasma 10. expand

*Label*

11. liquid 12. gas 13. solid

*Identify*

14. gas 15. liquid 16. solid

**Knowledge Builder (page 28)**

**Activity #1: Plastic Grocery Bag**

*Observation:* Answers will vary but should include the bag filled up with air.

*Conclusion:* Air takes up space. Gases in the air expanded to fill the plastic bag and took the shape of the bag.

**Activity #2: Balloon in a Bottle**

*Observation:* The balloon will not inflate.

*Observation:* The balloon inflated.

*Conclusion:* The bottle was full of air. The balloon did not have room to inflate. Once the air escaped through the hole placed in the bottle, there was room for the balloon to expand.

**Inquiry Investigation: Molecular Model (page 29)**

*Conclusion:*

**Solids:** have a definite shape and volume; keep their shape and take up the same amount of space; have a high cohesive force so the molecules of solids are packed tightly together; molecules are in constant motion; molecules are moving so slowly you cannot see them move

**Liquids:** have a definite volume but no definite shape; take up the same amount of space but will take the shape of the container in which they are placed; molecules have a low cohesive force; molecules spread far enough apart that they can flow over each other and move a little faster

**Gases:** have no definite volume or shape; cohesive force is almost nonexistent, so the molecules are not attracted to each other; the gases will expand to fill any container and will take the shape of the container

***Physical Properties of Matter***
**Quick Check (page 33)**
*Matching*
1. d 2. e 3. a 4. b 5. c

*Fill in the Blanks*
6. displacement 7. meniscus 8. Density
9. sink 10. sink, float

*Math Calculations*
11. 14,625 $cm^3$ 12. 945 $cm^3$

*Reading a Graduated Cylinder*
13. 52 mL 14. 37 mL 15. 65 mL

**Knowledge Builder (page 34)**
**Activity #2: Make Water Denser**
*Conclusion:* Adding salt to water makes the water denser. The water with the most salt will sink to the bottom of the straw.

**Inquiry Investigation: Density (page 35–36)**
*Hypothesis:* The more dense an object, the less it floats.
*Conclusion:* The objects with a density less than the density of the water floated. The objects with a density greater than the water sunk. My data supported/did not support my hypothesis that density affects an object's ability to float.

***Understanding Solids, Liquids, and Gases***
**Quick Check (page 40)**
*Matching*
1. c 2. b 3. d 4. a 5. e

*Fill in the Blanks*
6. characteristics 7. solid 8. elasticity
9. adhesion 10. Surface tension
11. cohesion
12. Buoyant force or Archimedes' Principle
13. Bernoulli's Principle

*Label*
14. surface tension 15. capillarity
16. elasticity 17. adhesion

**Knowledge Builder (page 41)**
**Activity #1: Capillary Action in Plants**
*Conclusion:* The flower absorbed colored water molecules from each beaker. The colored water moved up the plant stem to the flower. This illustrates capillarity.
**Activity #2: Viscosity**
*Conclusion:* Viscosity is a substance's resistance to flow. The slowest moving marble is in the most viscous liquid, the corn syrup.

**Inquiry Investigation: Elasticity (page 42–43)**
*Hypothesis:* The temperature of a rubber band affects/does not affect the distance it will stretch.
*Conclusion:* The temperature of a rubber band affects the distance it will stretch. The room-temperature rubber bands were most elastic and stretched farther than the frozen rubber bands before breaking. The data supported/did not support my hypothesis.

***Physical Changes of Matter***
**Quick Check (page 47)**
*Matching*
1. e 2. a 3. b 4. c 5. d

*Fill in the Blanks*
6. solid, liquid, gas, plasma 7. physical
8. cooled 9. Sublimation 10. diffusion

*Multiple Choice*
11. a 12. c 13. d 14. b 15. b

**Knowledge Builder (page 48)**
**Activity #1: Effects of Cooling and Heating Air**
*Conclusion:* Heating air causes the molecules to move faster and faster and farther apart. As they move farther apart, the molecules move out of the bottle and into the balloon. The molecules push on the sides of the balloon, causing it to inflate.
**Activity #2: Diffusion**
*Conclusion:* The temperature of the water determines how fast the diffusion takes place. The tea bag has a high concentration of tea; the water has a low concentration of tea. The tea molecules then have a tendency to move into the water until there is an equal concentration in each.

**Inquiry Investigation: Changing States of Matter (page 49–50)**
*Hypothesis:* Increased temperature will/will not cause ice to change states.
*Conclusion:* As a solid is heated, the molecules move faster and faster until the bonds weaken and the solid melts (changes to a liquid). The temperature at which a substance changes from a solid to a liquid is called the melting point. As the water is heated further, the molecules move even faster, and the liquid changes to a gas (steam). The temperature at which a substance changes from a liquid to a gas is called the boiling point. The data supports/does not support my hypothesis.

**Chemical Changes in Matter**
**Quick Check (page 54)**

*Matching*

1. b 2. d 3. a 4. c 5. e

*Fill in the Blanks*

6. chemical change
7. chemical bonds
8. chemical formula
9. pH scale
10. Litmus paper

*Data Table*

11. chemical formula
12. chemical equation
13. chemical equation
14. chemical formula

*Multiple Choice*

16. c 17. a

**Knowledge Builder (page 55)**
**Activity #1: pH Scale**

1. acid
2. neutral (slight acid)
3. base
4. acid
5. acid
6. base
7. acid
8. base
9. neutral (slight base)

**Activity #2: Chemical Formulas**

1. NaOH
2. $CO_2$
3. NaCl
4. $CaCO_3$
5. $KNO_3$

**Inquiry Investigation: Acids and Bases (page 56)**

*Results:*

1. acid
2. base
3. neutral
4. acid
5. acid
6. base
7. acid
8. base
9. neutral

*Conclusion:*

1. lemon juice
2. ammonia
3. It would be neutralized.

**Evidence of Chemical Changes**
**Quick Check (page 59)**

*Matching*

1. b 2. d 3. e 4. a 5. c

*Fill in the Blanks*

6. water
7. oxidation
8. Heat
9. endothermic
10. exothermic

*Data Table*

11. physical
12. chemical
13. chemical
14. chemical
15. chemical
16. chemical
17. physical
18. physical

*Identify*

19. endothermic
20. exothermic

**Knowledge Builder (page 60)**
**Activity #1: Endothermic and Exothermic Reactions**

1. exothermic, heat released
2. exothermic, heat released
3. exothermic, heat released
4. endothermic, heat absorbed
5. exothermic, heat released

**Activity #2: Physical Changes in Matter**

2. It is a physical change. Matter changed states, but no new substance was formed during the change. The molecules in the popcorn kernel start moving faster and faster when heated. The solid substance inside the kernel becomes a liquid. The liquid inside the kernel gets hotter, and the kernel starts to vibrate from the energy of the molecules bumping into each other inside the kernel. Finally, when the molecules of the liquid and the water inside get so hot they no longer can be contained in the kernel, they start popping out of the shell.

**Inquiry Investigation: Oxidation (page 61–62)**

*Hypothesis:* Oxidation affects/does not affect the properties of metal.

*Analysis:*

1. A chemical reaction occurred.
2. A chemical reaction occurred.
3. A new substance was formed that was not attracted to a magnet.

*Conclusion:* This is an example of a chemical reaction, because new substances are formed with different chemical and physical properties. In the first reaction, the steel wool changes color and an orange precipitate is formed, which is not attracted to a magnet. This orange powder is rust. In the second reaction, the orange powder changed to a blue-black color, and the substance is attracted to a magnet.

# Bibliography

**Children's Literature Resources:**

Bowden, M. (1997). *Chemical Achievers: The Human Face of the Chemical Sciences.* Philadelphia, PA: Chemical Heritage Foundation Publication.

Bowden, M. (1997). *Chemistry Is Electric.* Philadelphia, PA: Chemical Heritage Foundation Publication.

Chemical Heritage Foundation (Summer 2000). *Chemical Heritage,* 18(2). Philadelphia, PA: Chemical Heritage Foundation. *Chemical Heritage* is the journal for the Chemical Heritage Foundation.

Chemical Heritage Foundation (Fall 2000). *Chemical Heritage,* 18(3). Philadelphia, PA: Chemical Heritage Foundation.

Chemical Heritage Foundation (Winter 2000/1). *Chemical Heritage,* 18(4). Philadelphia, PA: Chemical Heritage Foundation.

Chemical Heritage Foundation (Spring 2001). *Chemical Heritage,* 19(1). Philadelphia, PA: Chemical Heritage Foundation.

Chemical Heritage Foundation (Summer 2001). *Chemical Heritage,* 19(2). Philadelphia, PA: Chemical Heritage Foundation.

Chemical Heritage Foundation (Fall 2001). *Chemical Heritage,* 19(3). Philadelphia, PA: Chemical Heritage Foundation.

Cooper, C. (1992). *Eyewitness Books: Matter.* London, England: Dorling Kindersley.

Feldman, A. and Ford, P. (1989). *Scientists and Inventors: The People Who Made Technology From Earliest Times to Present.* London: Godrey Cave Associates Limited.

Frese, G. (2000). *Dow Chemical Portrayed.* Philadelphia, PA: Chemical Heritage Foundation.

Hewitt, P., Suchocki, J., and Hewitt, L. (1999). *Conceptual Physical Science.* Menlo Park, CA: Addison Wesley Longman.

Hellemans, A. and Bunch, B. (1988). *The Timetables of Science: A Chronology of the Most Important People and Events in the History of Science.* New York, NY: Simon and Schuster.

Joly, D. (1988). *Grains of Salt.* Ossining, NY: Young Discovery Library.

Lindley, E. (1996). *Chemistry: Common Misconceptions and Fairy Tales.* Available online at: http://people.we.mediaone.net/elindley/commiscn.htm

Newman, A. (1993). *Eyewitness Books: Chemistry.* London, England: Dorling Kindersley.

Rayner-Canham, M. and Rayner-Canham, G. (1997). *A Devotion to Their Science: Pioneer Women of Radioactivity.* Philadelphia, PA: Chemical Heritage Foundation.

Rayner-Canham, M. and Rayner-Canham, G (1997). *Women in Chemistry: Their Changing Roles From Alchemical Times to the Twentieth Century.* Philadelphia, PA: Chemical Heritage Foundation.

**Children's Literature Resources (cont.):**

Rogers, K., Howell, L., Smith, A., Clarke, P., and Hederson, C. (2000). *The Usborne Internet-linked Science Encyclopedia*. London, England: Usborne Publishing Ltd.

University of Toronto (1998). *References for Misconceptions in Chemistry*. Available online at: http://www.oise.utoronto.ca/~science/chemmisc.htm

Wieland, P., ed. (1998). *Introducing the Chemical Science: A CHF Reading List*. Philadelphia, PA: Chemical Heritage Foundation.

Wertheim, J., Oxlade, C., and Stockley, C. (2000). *The Usborne Illustrated Dictionary of Chemistry*. Tulsa, OK: EDC Publishing.

Wilbraham, A., Staley, D., Matta, M., and Waterman, E. (2002). *Chemistry*. Glenview, IL: Addison Wesley.

Williams, T. (1987). *A History of Invention: From Stone Axes to Silicon Chips*. New York, NY: Facts on File Publications.

## Software:

Dorling Kindersley. (1995) *Encyclopedia of Science*. New York, NY: Dorling Kindersley Multimedia.

Houghton Mifflin. (1997) *Inventor Labs: Technology*. Pleasantville, NY. Houghton Mifflin Interactive.

Houghton Mifflin. (1997) *Inventor Labs: Transportation*. Pleasantville, NY. Houghton Mifflin Interactive.

Microtel (1999). *The Science Club: Just a Chemical Reaction*. Montreal, Canada: Microtel Inc.

Stranger, D. (pub). (1998). *Thinkin' Science Series: ZAP!*. Redmond, WA: Edmark Corporation.

## Websites:

www.brainpop.com/science/seeall.wem
www.chem4kids.com/index.html
www.chemicalelements.com/
www.exploratorium.edu/science_explorer/
www.funbrain.com/periodic/index.html
www.howstuffworks.com/
www.iop.org/Physics/Electron/Exhibition/
www.its.caltech.edu/~atomic/snowcrystals
www.mathmol.com/textbook/middle_home.html
www.miamisci.org/af/sln/phantom/
www.miamisci.org/af/sln/phases
www.mos.org/sln/Leonardo
www.pbs.org/wgbh/aso/tryit/atom/
www.pbs.org/wgbh/nova/lostempires/obelisk/
www.usborne-quicklinks.com

## Curriculum Resources:

American Institute of Physics (2000). *Children's Misconceptions About Science.* American Institute of Physics. Available online at: http://www.amasci.com/miscon/opphys.html.

American Institute of Physics (Circa 1988) *Operation Physics: Matter and Its Changes*. American Institute of Physics.

Barber, J. (1986).*Chemical Reactions.* Berkeley, CA: Lawrence Hall of Science.

Chemical Education for Public Understanding Program. (Circa 1988). *Chemicals, Health, Education and Me.* Berkeley, CA: Lawrence Hall of Science.

Chemical Education for Public Understanding Program. (Circa 1988). *Chemical Education for Public Understanding Program*. Berkeley, CA: Lawrence Hall of Science.

Gertz, S., Portman, D, and Sarquis, M. (1996). *Teaching Physical Science Through Children's Literature: 20 Complete Lessons for Elementary Grades*. Middletown, OH: Terrific Science Press.

Glover, D. (1993). *Solids and Liquids.* New York, NY: Kingfisher.

Greenberg, B. and Patterson, D. (1998). *Art in Chemistry: Chemistry in Art*. Englewood, CO: Teacher Ideas Press.

Liem, T. (1992). *Invitations to Science Inquiry: Over 400 Discrepant Events to Interest and Motivate Your Students in Learning Science*. Chion Hills, CA: Science Inquiry Enterprises.

Lorbeer, G. (2000). *Science Activities for Middle School Students*. Boston, MA: McGraw-Hill.

Marsland, D. (2000). *Science and Technology Concepts for Middle Schools: Properties of Matter*. Burlington, NC: Carolina Biological Supply Company.

Sarquis, M. (1997). *Exploring Matter With Toys: Using and Understanding the Senses.* Middletown, OH: Terrific Science Press.

Sarquis, J., Hogue, L., Sarquis, M., and Woodward, L. (1997). *Investigating Solids, Liquids, and Gases With Toys: States of Matter and Changes of State.* Middletown, OH: Terrific Science Press.

Sarquis, M. and Sarquis, J. (1991). *Fun With Chemistry: A Guidebook of K-12 Activities, Volume One.* Madison, WI: Institute for Chemical Education.

Sarquis, J., Sarquis, M., and Williams, J. (1995). *Teaching Chemistry With Toys: Activities for Grades K-9.* Middletown, OH: Terrific Science Press.

Sherwood, M., (ed.) (1985). *Chemistry Today*. Chicago, IL: World Book Encyclopedia.

Taylor, B. (1998). *Teaching Energy With Toys: Complete Lessons for Grades 4-8*. Middletown, OH: Terrific Science Press.